Stretch Your Thinking

ENRICHMENT WORKBOOK

Harcourt Brace & Company

Orlando • Atlanta • Austin • Boston • San Francisco • Chicago • Dallas • New York • Toronto • London

http://www.hbschool.com

CONTENTS

Twelve Holiday Presents

In a traditional holiday song, a present is given each day for
12 days.

Use a calculator to help you complete the chart.

1. Multiply to find the total cost of each present. For example,
   a partridge in a pear tree is given every day for twelve days.
   Multiply $27.50 × 12.

2. Add to find the total cost of all the presents.

Day	Present	Price	Multiply	Total Cost
First	A partridge in a pear tree	$27.50	× 12	
Second	2 turtle doves	$50.00	× 11	
Third	3 French hens	$15.00	× 10	
Fourth	4 calling birds	$280.00	× 9	
Fifth	5 golden rings	$325.00	× 8	
Sixth	6 geese-a-laying	$150.00	× 7	
Seventh	7 swans-a-swimming	$3,500.00	× 6	
Eighth	8 maids-a-milking	$38.00	× 5	
Ninth	9 ladies dancing	$3,258.54	× 4	
Tenth	10 lords-a-leaping	$3,182.57	× 3	
Eleventh	11 pipers piping	$1,137.24	× 2	
Twelfth	12 drummers drumming	$1,232.01	× 1	
Total				

Benchmarking in Sports

The concept of benchmarking is used by many athletes in timed distance events: running, biking, skiing, and swimming.

You will need: a calculator.

For a runner in a mile race, the benchmark is usually one-quarter mile—one lap of a standard track. A runner who wants to run a 4-minute mile needs to run each lap in 60 seconds. Complete the chart for the mile run.

	One-Mile Run	
	Desired Time	**Benchmark Time (per lap)**
1.	5-minute mile	75 seconds
2.		90 seconds
3.	7-minute mile	
4.		120 seconds

In a 10-mile race, a runner usually uses a benchmark of one mile. A runner who wants to run 10 miles in 50 minutes needs to run a 5-minute mile. Complete the chart for the 10-mile run.

	Ten-Mile Run	
	Desired Time	**Benchmark Time (per lap)**
	50 minutes	5 minutes
5.		4 minutes, 45 seconds
6.	55 minutes	
7.		6 minutes
8.	62 minutes, 30 seconds	
9.		7 minutes
10.	1 hour, 20 minutes	

The Chinese Abacus

An abacus is one of the first calculating tools. The best known
Chinese abacus, called the *suan pan,* is shown in this lesson.

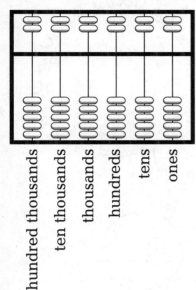

hundred thousands · ten thousands · thousands · hundreds · tens · ones

← Each bead above the bar represents 5 units.

← center bar

← Each bead below the bar represents 1 unit.

← The abacus is arranged like a place-value chart.

To make and read numbers, you move beads toward the
center bar and add. For example:

4 7 0 9 5 = 47,095

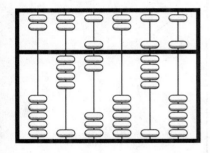

2 3 8 4 6 1 = 238,461

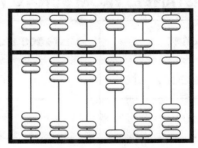

Write the number shown on the abacus.

1.

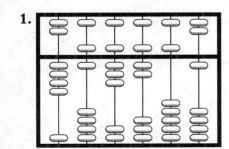

_____

2.

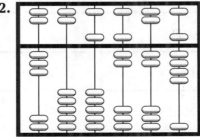

_____

Some Solar Sums

The table lists the approximate distances between planets.

DISTANCES BETWEEN PLANETS	
Planets	**Distance Between (in miles)**
Sun to Mercury	36,000,000 mi
Mercury to Venus	31,000,000 mi
Venus to Earth	26,000,000 mi
Earth to Mars	48,600,000 mi
Mars to Jupiter	341,700,000 mi
Jupiter to Saturn	403,100,000 mi
Saturn to Uranus	899,600,000 mi
Uranus to Neptune	1,008,000,000 mi
Neptune to Pluto	866,000,000 mi

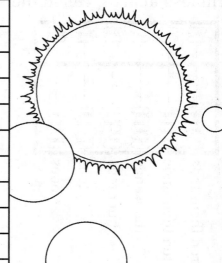

Use the information in the table above to write how many miles each planet is from the sun.

	DISTANCES FROM SUN TO PLANETS	
	Planet	**Distance from Sun (in miles)**
1.	Mercury	
2.	Venus	
3.	Earth	
4.	Mars	
5.	Jupiter	
6.	Saturn	
7.	Uranus	
8.	Neptune	
9.	Pluto	

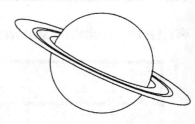

Popular Hot Spots

Listed below are the populations for major cities
in Florida and Texas. For each state, order the
cities from largest population to smallest
population. Write the rank on the line.

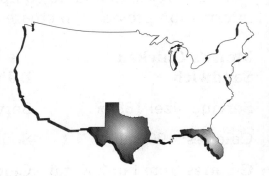

Florida Cities	City Population	Rank
Fort Lauderdale	149,377	
Hialeah	188,004	
Jacksonville	635,230	1
Miami	358,548	2
Orlando	164,693	
Saint Petersburg	238,629	
Tallahassee	124,773	
Tampa	280,015	

Texas Cities	City Population	Rank
Austin	465,622	
Beaumont	114,323	
Brownsville	98,962	
Corpus Christi	257,453	
Dallas	1,006,877	
El Paso	515,342	
Fort Worth	447,619	
Houston	1,630,553	
Laredo	122,899	
Lubbock	186,206	
San Antonio	935,933	
Waco	103,590	
Wichita Falls	96,259	

Comparing Chicken Sandwiches

You can create a table to compare information. Use the
information provided to create a comparison table.

**Molly's Chicken
Sandwich**

Serving size: 189 g

Calories: 490 cal

Calories from fat: 260 cal

Cholesterol: 50 mg

Sodium: 800 mg

B.J's Chicken Sandwich

Serving Size: 189 g

Sodium: 1,155 mg

Calories from fat: 314 cal

Calories: 578 cal

Cholesterol: 50 mg

**Wimpy's Chicken
Sandwich**

Serving Size: 189 g

Cholesterol: 35 mg

Sodium: 763 mg

Calories: 399 cal

Calories from fat: 145 cal

Use the table to compare. Answer each question.

1. List the sandwiches from least to greatest based on the
   number of calories.

   _____

2. List the sandwiches from least to greatest based on the
   amount of sodium.

   _____

3. The total calories for each sandwich are shown. Which
   sandwiches get more than half their calories from fat?

   _____

Addition Squares

In the squares below, the total is always three times the value
of the number in the center of the box. Each horizontal,
vertical, and diagonal row must add up to the same total.

11	13	
	10	

11	13	6
5	10	15
14	7	9

The total for this square is 30.

For example, 6 + 15 + 9 = 30

Complete each square. Use any numbers from 0 to 15. You may
use a number up to three times within each magic square.

1.

6		
	5	
9		

Total = 15

2.

6		
4	5	

Total = 15

3.

5		
10	6	

Total = 18

4.

6		
	6	
	5	

Total = 18

5.

11	8	
	7	

Total = 24

6.

15	8	
	9	

Total = 24

Mystery Message

Why did the astronaut throw a plate out the window?

To find out, subtract.

1. 2,632
    − 1,247
          T

2. 463
    − 256
          G

3. 583
    − 477
          L

4. 826
    − 538
          O

5. 744
    − 369
          U

6. 987
    − 628
          A

7. 1,465
    − 876
          F

8. 475
    − 366
          I

9. 4,593
    − 2,646
          Y

10. 8,251
     − 4,627
           E

11. 6,976
     − 2,697
           N

12. 7,224
     − 3,568
           C

13. 4,280
     − 1,699
           S

14. 529
     − 146
           R

15. 8,951
     − 1,598
           !

Locate each of your answers below. Put the letter that corresponds to it in the matching box to discover the answer to the riddle.

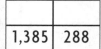

1,385	288

2,581	3,624	3,624

359

589	106	1,947	109	4,279	207

2,581	359	375	3,656	3,624	383	7,353

Daily Cross-Number Puzzle

Subtract. Enter your answers in the cross-number puzzle.

Across

1. $\begin{array}{r} 300 \\ -\ 158 \\ \hline \end{array}$

4. $\begin{array}{r} 284 \\ -\ 102 \\ \hline \end{array}$

7. $\begin{array}{r} 2,000 \\ -\ 1,177 \\ \hline \end{array}$

8. $\begin{array}{r} 800 \\ -\ 685 \\ \hline \end{array}$

9. $\begin{array}{r} 1,400 \\ -\ 1,113 \\ \hline \end{array}$

10. $\begin{array}{r} 10,000 \\ -\ 9,925 \\ \hline \end{array}$

11. $\begin{array}{r} 5,001 \\ -\ 2,438 \\ \hline \end{array}$

12. $\begin{array}{r} 1,710 \\ -\ 189 \\ \hline \end{array}$

13. $\begin{array}{r} 10,201 \\ -\ 2,238 \\ \hline \end{array}$

14. $\begin{array}{r} 9,007 \\ -\ 4,789 \\ \hline \end{array}$

Down

1. $\begin{array}{r} 3,008 \\ -\ 1,191 \\ \hline \end{array}$

2. $\begin{array}{r} 5,200 \\ -\ 985 \\ \hline \end{array}$

3. $\begin{array}{r} 700 \\ -\ 465 \\ \hline \end{array}$

4. $\begin{array}{r} 25,000 \\ -\ 12,245 \\ \hline \end{array}$

5. $\begin{array}{r} 1,280 \\ -\ 1,192 \\ \hline \end{array}$

6. $\begin{array}{r} 1,000 \\ -\ 973 \\ \hline \end{array}$

7. $\begin{array}{r} 6,000 \\ -\ 5,183 \\ \hline \end{array}$

9. $\begin{array}{r} 8,000 \\ -\ 5,245 \\ \hline \end{array}$

12. $\begin{array}{r} 400 \\ -\ 262 \\ \hline \end{array}$

13. $\begin{array}{r} 104 \\ -\ 30 \\ \hline \end{array}$

15. $\begin{array}{r} 10,008 \\ -\ 3,808 \\ \hline \end{array}$

16. $\begin{array}{r} 8,900 \\ -\ 5,800 \\ \hline \end{array}$

Number Pyramids

Number pyramids gain new levels by adding together the level beneath. Use this simple pattern:

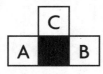

A + B = C

For example, given 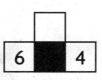 6 + 4 = 10. So,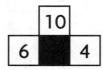

Depending on which numbers are given, you may also use subtraction: C − B = A or C − A = B.

Solve the number pyramids.

1.

2.

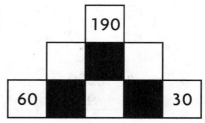

3.

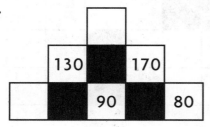

4.

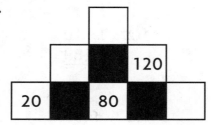

5.

6.

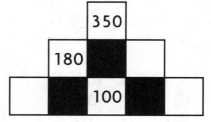

7. Make two of your own pyramids.

Name _____

Super (Market) Estimations

Cashiers can make errors, and scanners don't always scan the correct prices. It is important to check your receipt.

At the left is a list of your purchases. At the right is what the cash register rang up. Match the lists and circle the errors. By how much was the receipt off?

		Market Receipt
Facial tissues	$1.29	4.50
Fruit drink	$1.79	1.96
Rice	$1.69	0.65
Soap	$0.89	1.99
Apples—3 lbs. at	$1.50 lb.	2.98
Light bulbs	$2.89	0.97
Carrots	$0.65	1.29
Cereal	$3.49	3.49
Milk	$1.39	4.39
Butter	$1.99	8.90
Sugar	$0.79	1.56
Flour	$0.75	1.79
Soda	$3.49	0.30
Oatmeal	$1.56	1.39
Bagels	$3.00	0.75
Bread	$1.59	4.79
Mustard	$3.10	2.75
Cookies	$2.75	3.10
Chicken	$4.97	1.59

Total _____   **Total** _____

The receipt was off by _____ .

Shop Till You Drop!

Estimate the cost of the items on each list. Circle the list that comes closest without going over your spending limit.

1. Your spending limit is $400.

List 1
Suit
Shirt
Shoes
Coat
Gloves

Suit	$185	Belt	$32
Shirt	$36	Coat	$115
Shoes	$44	Hat	$46
Tie	$28	Pants	$29
Gloves	$13	Suspenders	$35
Socks	$7		

List 2
Coat
Hat
Shirt
Suit
Belt

Estimated cost: _____

Actual cost: _____

Estimated cost: _____

Actual cost: _____

2. Your spending limit is $2,000.

List 1
Computer
CD-ROM drive
Printer
Software
Speakers

Computer	$1,200	Joystick	$59
Laptop Computer	$1,500	Desk	$79
CD-ROM drive	$239	Speakers	$139
Printer	$319		
Software	$179		

List 2
Laptop Computer
Printer
Software

Estimated cost: _____

Actual cost: _____

Estimated cost: _____

Actual cost: _____

About the Abacus

The abacus is a calculating machine that has been used for thousands of years. In ancient times, the abacus consisted of rows of grooves in sand into which pebbles were placed. Today's abacus, known as the Chinese *suan pan*, consists of beads strung on parallel wires. Each wire represents a place value.

Each bead above the bar represents 5.
The beads move down to count 5s.

Each bead below the bar represents 1.
The beads move up to count 1s.

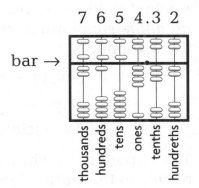

7 6 5 4.3 2

bar →

thousands hundreds tens ones tenths hundreths

Write the number represented by the abacus.

1. _____

2. _____

3. _____

4. _____

5. _____

6. _____

Draw beads on the abacus to represent the number.

7. 2358.41

8. 8642.09

9. 3962.15

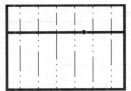

Take Note

A sheet of paper is only about 0.0032 inch thick. In *scientific notation*, small numbers such as this are written using negative exponents.

$$0.0032 = 3.2 \times 10^{-3} \longleftarrow \text{exponent}$$

The decimal point moves 3 places to the right.

The exponent $^-3$ shows how many places the decimal point moves to the right. Here are more examples.

2.5	=	2.5×10^{0}	$\longleftarrow$	The exponent 0 means the decimal point does not move to the right.
0.25	=	2.5×10^{-1}	$\longleftarrow$	The exponent $^-1$ means the decimal point moves 1 place to the right.
0.025	=	2.5×10^{-2}	$\longleftarrow$	The exponent $^-2$ means the decimal point moves 2 places to the right.
0.0025	=	2.5×10^{-3}	$\longleftarrow$	The exponent $^-3$ means the decimal point moves 3 places to the right.

Fill in the box with the correct negative exponent.

1. $0.014 = 1.4 \times 10^{\boxed{}}$

2. $0.00051 = 5.1 \times 10^{\boxed{}}$

3. $0.021 = 2.1 \times 10^{\boxed{}}$

4. $0.00000086 = 8.6 \times 10^{\boxed{}}$

5. $.000033 = 3.3 \times 10^{\boxed{}}$

6. $0.000000000074 = 7.4 \times 10^{\boxed{}}$

Write each number in scientific notation.

7. $0.0036 =$ _____ $\times$ _____

8. $0.71 =$ _____

9. $0.043 =$ _____

10. $0.000000052 =$ _____

11. $0.000035 =$ _____

12. $0.00000000027 =$ _____

13. A red blood cell is about 0.0000072 m wide. _____

14. A platelet is about 0.000003 m wide. _____

Name _____

Place the Values

You can make several different numbers that all use the same digits.

The following numbers all use the digits 0, 2, 5, 6, and 9.

205.69, 506.92, 592.06, 620.59

Complete the tables. There are many different correct answers.

1. Use the numbers 0, 1, 2, 5, 6, 7, 9. Create four 7-digit numbers of increasing value.

Thousands	Hundreds	Tens	Ones	Tenths	Hundredths	Thousandths

2. Use the numbers 0, 2, 3, 4, 6, 8, 9. Create four 7-digit numbers of decreasing value.

Thousands	Hundreds	Tens	Ones	Tenths	Hundredths	Thousandths

3. Use the numbers 0, 1, 2, 4, 5, 7, 8. Create four 7-digit numbers of increasing value.

Thousands	Hundreds	Tens	Ones	Tenths	Hundredths	Thousandths

The Equalizer

Each number in the column at the right is a decimal equal to a number in a circle. Write the letter of the decimal next to its equivalent in a circle. You may use some letters more than once.

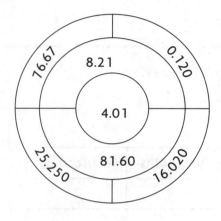

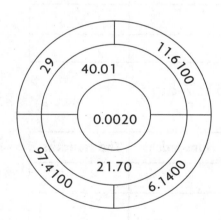

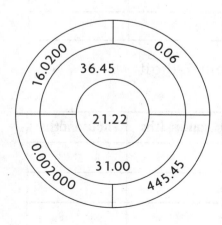

A	=	0.002
B	=	16.02
C	=	40.0100
D	=	36.450
E	=	0.060
F	=	81.6
G	=	6.14
H	=	31.0
I	=	21.220
J	=	29.000
K	=	11.61
L	=	25.25
M	=	445.450
N	=	21.7
O	=	76.67000
P	=	8.210
Q	=	4.0100
R	=	97.41
S	=	0.12
T	=	0.00200
U	=	16.0200

Make three words using the letters in each circle.

_____ _____ _____

Swim To Win

You are an official at a swim meet. You recorded the times of the swimmers. Now you need to rank the times from fastest to slowest so that everyone knows in what place the swimmers finished.

Complete the charts.

100-Meter Freestyle

Lane 1   48.59

Lane 2   49.01

Lane 3   48.45

Lane 4   48.95

Lane 5   49.10

Lane 6   48.54

Lane 7   48.99

Place	Lane	Time (sec)
1		
2		
3		
4		
5		
6		
7		

100-Meter Butterfly

Lane 1   53.32

Lane 2   53.16

Lane 3   53.19

Lane 4   53.23

Lane 5   53.14

Lane 6   53.56

Lane 7   53.41

Place	Lane	Time (sec)
1		
2		
3		
4		
5		
6		
7		

100-Meter Backstroke

Lane 1   55.13

Lane 2   54.55

Lane 3   54.39

Lane 4   54.60

Lane 5   54.75

Lane 6   55.03

Lane 7   54.31

Place	Lane	Time (sec)
1		
2		
3		
4		
5		
6		
7		

Fast, Faster, Fastest

OFFICIAL WORLD LAND SPEED RECORDS			
Date	**Driver**	**Car**	**mph**
Sep 3, 1935	M. Campbell	Bluebird Special	301.13
Nov 19, 1937	Eyston	Thunderbolt #1	311.42
Aug 27, 1938	Eyston	Thunderbolt #1	345.5
Sep 15, 1938	Cobb	Railton	350.2
Sep 16, 1938	Eyston	Thunderbolt #1	375.5
Aug 23, 1939	Cobb	Railton Red Lion	368.9
Sep 16, 1947	Cobb	Railton Mobil Special	394.19
Jul 17, 1964	D. Campbell	Bluebird	403.1
Oct 2, 1964	Green	Winged Foot Express	413.20
Oct 5, 1964	Arfons	Green Monster	434.02
Oct 13, 1964	Breedlove	Spirit of America	468.72
Oct 15, 1964	Breedlove	Spirit of America	526.28
Oct 27, 1964	Arfons	Green Monster	536.71
Nov 2, 1965	Breedlove	Spirit of America-Sonic One	555.13
Nov 7, 1965	Arfons	Green Monster	576.55
Nov 15, 1965	Breedlove	Spirit of America-Sonic One	600.60
Oct 23, 1970	Gabelich	Blue Flame	622.407

Source: Grolier's Multimedia Encyclopedia, 1995

1. Which car won the most races?

   _____

2. List Breedlove's speeds from least to greatest.

   _____

3. List the speeds of Railton, Winged Foot Express, Bluebird Special, and Blue Flame from greatest to least.

   _____

T's Anyone?

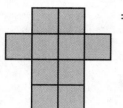 = 1.0   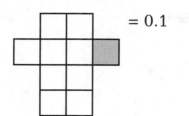 = 0.1

Shade the figures to represent each decimal. Find the sum and shade the figure.

1.   1.3   +   1.2   =   ____

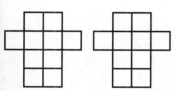   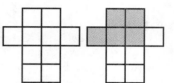

2.   1.8   +   0.6   =   ____

3.   2.4   +   1.7   =   ____

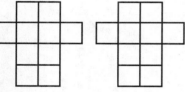 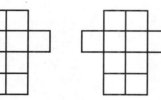 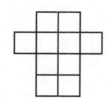 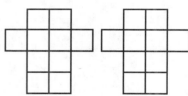

4.   3.5

2.9

     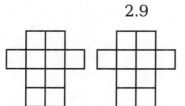 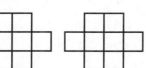   =   ____

How Much Farther?

Freddy Four is traveling from Count City to Problem Solver's Landing. He knows the distance between Problem Solver's Landing and Count City is 297 miles.

The distance between each town is marked on the map. At each town, write down the distance remaining for Freddy to travel.

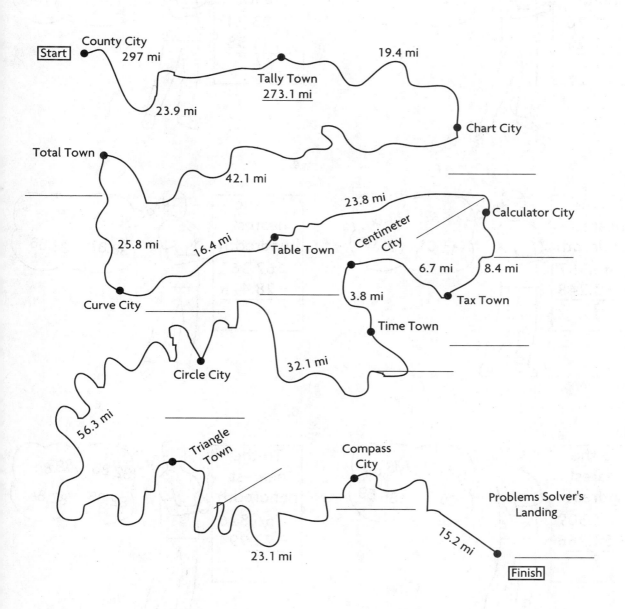

Start — County City   297 mi

Tally Town
273.1 mi

19.4 mi

23.9 mi

Chart City

Total Town

42.1 mi

23.8 mi   Calculator City

25.8 mi

16.4 mi   Table Town   Centimeter City

6.7 mi   8.4 mi

Curve City

3.8 mi

Tax Town

Time Town

Circle City

32.1 mi

56.3 mi

Triangle Town

Compass City

Problems Solver's Landing

15.2 mi

23.1 mi

Finish

Estimate

Round each number to the place value noted on the card. Then add or subtract. Circle the estimate for each problem.

1.

To the nearest tenth

29.76
+37.21
?

67   66.9
67.1   66.8

2.

To the nearest tenth

23.24
−11.58
?

11.7   11.5
11.4   11.6

3.

To the nearest hundredth

48.357
+63.248
?

111.59   111.6
111.61   111.65

4.

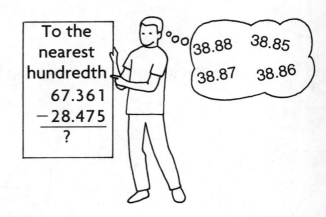

To the nearest hundredth

67.361
−28.475
?

38.88   38.85
38.87   38.86

5.

To the nearest hundredth

84.309
+51.266
?

135.57   135.56
135.58   135.59

6.

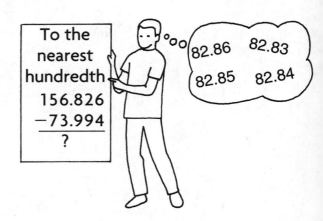

To the nearest hundredth

156.826
−73.994
?

82.86   82.83
82.85   82.84

Around the World

In each box write a number to be added or subtracted. Write each answer on the line with the equals sign. Begin at start. Add or subtract clockwise. Be sure that your answer at the end equals the starting number.

1.

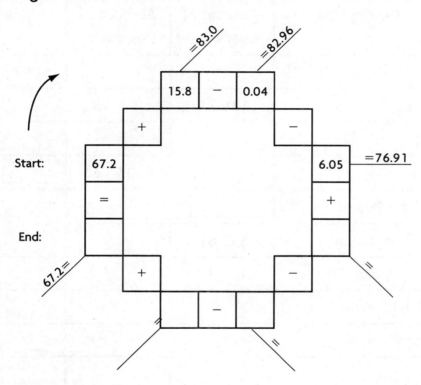

2.

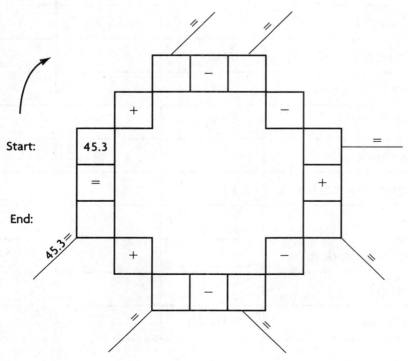

Name _____

My Balance!

Ted forgot to enter all of his checks and deposits into his check register. Fill in the missing information to help Ted find the balance in his account.

Check Number	Date	Description	Amount of Check	Amount of Deposit	Balance
					$897.54
645	1/17	Shirts Galore	$38.75		_____
646	1/18	Newton News	_____		_____
	1/18	paycheck		$325.76	_____
647	1/18	Burger Buster	_____		_____
648	1/19	Snipper Salon	$13.67		_____
649	1/20	Ring-A-Ling	$144.91		_____
650	1/20	Walkin' Wear	_____		_____
651	1/20	Harry's Hats	$478.23		_____
652	1/21	Auto Al	_____		_____
	1/21	bonus check		$675.25	_____
653	1/21	Cartoon Car Wash	$40.70		_____
654	1/25	Karen's Kitchen	_____		_____
655	1/25	Pool	$98.04		_____
	1/25	paycheck		_____	_____
656	1/28	Gregg's Groceries	$123.32		_____
657	1/28	vet	_____		_____
658	1/29	Farmer's Market	_____		_____
659	1/29	Flowers	$187.50		_____
660	1/29	bank fee	$10.50		_____

Flying Away

Sharon has let go of her balloons. The only way they won't fly away is if you solve the equations. Find each product. Then use the directions below to color the balloons.

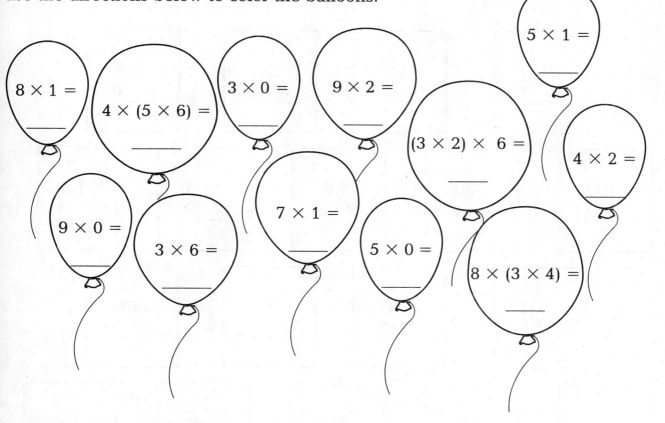

1. Color red the balloons whose product is zero.

2. Color blue the balloons that have a factor of 1.

3. Color green the balloons that have 3 factors that can be grouped differently.

4. Color orange the balloons that are left. Write something about the factors in these balloons.

_____

_____

5. Why do the red balloons equal 0?

_____

Moving Day

The Barretts are moving. They have color coded the boxes by how many groups of items are in each box. Solve the problems. Use the table below to color code the Barretts' boxes.

| 316 × 3 | 428 × 2 | 581 × 4 | 603 × 6 |

| 337 × 2 | 501 × 2 | 729 × 3 | 257 × 4 | 147 × 5 |

| 932 × 6 | 369 × 3 | 421 × 4 | 504 × 3 |

| 803 × 4 | 642 × 5 | 678 × 3 | 826 × 3 | 943 × 2 |

Groups	Colors
2	red
3	blue
4	orange
5	yellow
6	green

In what color box are most of the Barretts' items?

_____

Crossing Numbers

Find the products.

246 × 4 = _____   621

× _____   ×

3 × 157 = _____   4

= ×   =   2

7 × _____ = _____   2   _____   ×

×   =   476

361   3 × 314 = _____   7 × 425 = _____

=   ×   ×   _____

_____   401   6 × 555 = _____

=   =   ×

_____   _____   4 × 345 = _____

=   ×

8   2

×   =

229 × 7 = _____   _____

=

_____

See How Many You Can Make!

Find the volume of each prism. Then draw different figures that have the same volume. Be sure at least one side of each unit cube touches the side of another unit cube.

1.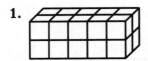

Volume = _____

2.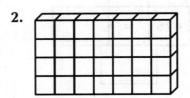

Volume = _____

3.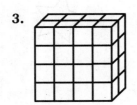

Volume = _____

Where Are They?

Find each word in the Word-Find from the list below. The words may be found across, down, diagonally, and backward.

c	e	n	t	i	m	e	t	e	r	s	h	a	i
p	r	e	a	p	e	m	u	l	o	v	v	b	n
f	c	h	e	i	g	h	t	s	h	a	p	e	s
o	i	n	r	h	v	i	h	t	g	n	e	l	t
r	b	i	a	l	g	n	u	m	b	e	r	s	u
m	u	l	t	i	p	l	i	c	a	t	i	o	n
u	c	e	d	i	s	q	u	a	r	e	m	f	i
l	s	e	j	a	m	c	i	n	c	h	e	s	t
a	t	s	r	e	t	e	m	h	a	p	t	q	s
d	e	c	i	m	a	l	s	z	z	g	e	k	o
u	s	d	r	a	y	t	r	e	p	o	r	p	v

volume	area	multiplication	inches
cubic	height	perimeter	square
units	length	meters	yards
centimeters	numbers	times	decimals
digit	property	shapes	formula

What is the area of the Word-Find if each square represents 1 inch?

_____

Formula Fun

Jodi's Super Store has purchased some new items.
She needs to find out the volume of each item for her advertising.

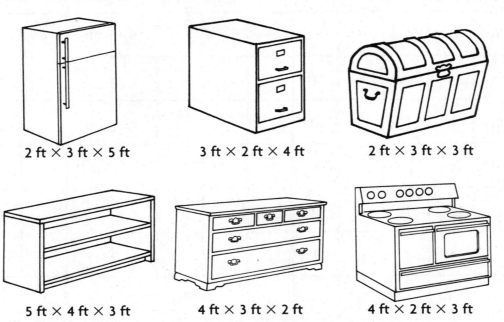

2 ft × 3 ft × 5 ft   3 ft × 2 ft × 4 ft   2 ft × 3 ft × 3 ft

5 ft × 4 ft × 3 ft   4 ft × 3 ft × 2 ft   4 ft × 2 ft × 3 ft

Complete the table. Fill in the name of each item that fits the
measurements. Then find the volumes of the items.

Item	Length	Width	Height	Volume
_____	2 ft	3 ft	5 ft	_____
_____	3 ft	2 ft	4 ft	_____
_____	5 ft	4 ft	3 ft	_____
_____	4 ft	3 ft	2 ft	_____
_____	4 ft	2 ft	3 ft	_____
_____	2 ft	3 ft	3 ft	_____

1. Which item has the greatest volume? the least volume?

2. Which items have the same
   volume?

_____   _____

Cross-Number Puzzle

A cross-number puzzle is a multiplication puzzle that works like the Distributive Property.

Solve the puzzle $9 \times 8 = n$ this way.
- Put the factors in the squares.
- Break each factor into 2 of its addends. Record the addends along the top and right side of the drawing.

$9 = 5 + 4$

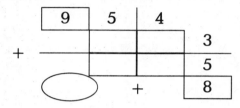

$8 = 5 + 3$

- Multiply the addends. Record the products in the inside boxes.
- Add the products horizontally and vertically.
- Record the sums along the bottom and left side of the drawing.
- Add the sums. The sum of the 2 numbers at the bottom should equal the sum of the 2 numbers on the left side.
- Put this number in the circle; this is the product of the original factors.

	9	5	4	
+	27	15	12	3
	45	25	20	5
	(72)	40 + 32		8

So, $9 \times 8 = 72$.

Complete the cross-number puzzles.

1. $8 \times 7 = n$

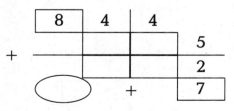

2. $14 \times 36 = n$

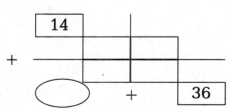

The Case of the Lost Digits

Some numbers are missing. Be a math detective and find the missing digits.

1.
```
        □   8
    ×   3   □
    ─────────
        □ □ 4
  + 1 , 7 4 0
  ───────────
    1 , 9 1 4
```

2.
```
        7   □
    ×   2   8
    ─────────
        6 □ □
  + 1 , 5 0 0
  ───────────
    2 , □ 0 0
```

3.
```
        □   7
    ×   4   □
    ─────────
        2   0 □
  + □ , □ 8 0
  ───────────
    2 , 8 8 1
```

4.
```
    2   3   8
    ×   □   □
    ─────────
  1 , 4 □ 8
  + 9 , □ 2 □
  ───────────
  1 0 , 9 4 □
```

5.
```
    4   6   7
    ×   □   8
    ─────────
  3 , 7 □ □
 + 1 4 , 0 1 □
 ───────────
 1 7 , 7 □ □
```

6.
```
    5   □   4
    ×   4   □
    ─────────
    □ , □ 3 8
 + 2 1 , 3 □ 0
 ───────────
 □ □ , 0 9 8
```

7.
```
    □ □ □
    ×   6   3
    ─────────
  1 , □ 0 8
 + 3 8 , 1 6 □
 ───────────
 4 0 , 0 6 8
```

8.
```
    7   □   5
    ×   3   □
    ─────────
    □ 2   5
 + 2 □ , 7 □ □
 ───────────
 2 2 , 4 7 5
```

9.
```
    7   5   □
    ×   □   5
    ─────────
  3 , □ 5 □
 + 3 □ , 0 0 □
 ───────────
 3 3 , 7 5 0
```

10.
```
    3   □   8
    ×   □   5
    ─────────
  1 , 8 □ □
 + 1 1 , □ 4 0
 ───────────
 □ □ , 2 3 0
```

11.
```
    6   2   6
    ×   □   □
    ─────────
  □ , □ 0 □
 + 1 2 , 5 □ 0
 ───────────
 □ □ , 0 2 4
```

12.
```
    □   5   □
    ×   1   □
    ─────────
  1 , 3 □ 1
 + 4 , □ 7 0
 ───────────
 5 , □ 4 1
```

Target Practice

Practice your estimation skills in this challenging game.
You will need a calculator.

The object of the game is to choose factors that produce
a product as close as possible to a chosen target.

Work with a partner to solve.

List A			List B		
473	698	5444	23	72	49
541	237	629	41	61	27
812	1010	303	18	36	54
349	421	568	32	15	45

Step 1 Choose a number from List A as your target. Circle it.

Step 2 Choose a number from List B, circle it, and enter it on
your calculator.

Step 3 Without estimating or doing mental math, enter ×,
then the first number that comes to mind, and then =.
Record your product.

Step 4 Repeat Steps 2–3 two more times, each time trying
to get closer to your target. Circle the product closest to
your target.

Now your partner uses the calculator and performs Steps 1–4,
using any numbers not circled in Lists A and B.

The player whose circled product is closer to his or her target
gets 1 point. The first player to reach 6 points wins.

Name _____ •_____

Letter Go!

Each letter stands for a 1-digit number. Find a value for each letter.

1.
```
  AAA
 +BBB
 ─────
  CCC
```
There are 32 possible solutions.

____ ____

2.
```
  MMM
  NNN
 +PPP
 ─────
  QQQ
```
There are 42 possible solutions.

____ ____

3.
```
  TTT
 ×   S
 ─────
  RRR
```
There are 4 possible solutions.

____ ____ ____ ____

4.
```
   JJJ
 ×  KK
 ─────
   JJJ
  JJJ
 ─────
  JLLJ
```
There are 3 possible solutions.

____ ____ ____ ____

____ ____ ____ ____

5.
```
    EEE
 ×  FFF
 ──────
    EEE
   EEE
  EEE
 ──────
  EGHGE
```
There are 2 possible solutions.

____ ____ ____

____ ____

6.
```
   XX
 × YY
 ────
   XX
  XX
 ────
  XZX
```
There are 2 possible solutions.

____ ____

____ ____

Some of the Parts

If you can find the perimeter and area of a rectangle, then you can find the perimeters and areas of these figures.

1.

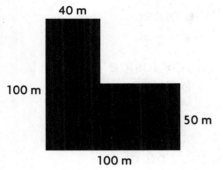

40 m

100 m

50 m

100 m

perimeter = _____

area   = _____

2.

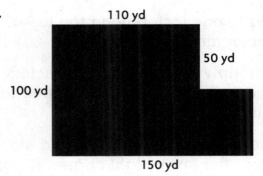

110 yd

50 yd

100 yd

150 yd

perimeter = _____

area   = _____

3.

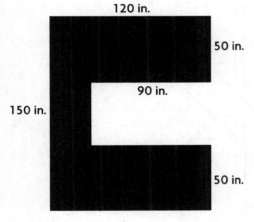

120 in.

50 in.

90 in.

150 in.

50 in.

perimeter = _____

area   = _____

4.

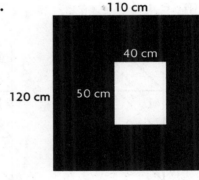

110 cm

40 cm

120 cm   50 cm

perimeter = _____

area   = _____

The area = the area of the large rectangle minus the area of the small rectangle.

Lattice Multiply

An early method of multiplying is the lattice method.
Here's how it works.

Multiply 342 × 657.

- Write one factor along the top of the lattice and the other
  factor along the right side.

- Multiply each digit of the factors. Record the products inside
  the lattice so that the ones and tens are separated by a
  diagonal. (See Figure 1.)

- Add the numbers in the grid along the diagonals, starting
  from the lower right corner. Record each sum at the end of
  its diagonal. Carry any tens into the next diagonal—just as
  you do when adding columns. (See Figure 2.)

- Read the digits down the left and across the bottom. This is
  the product.

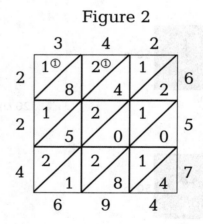

Figure 1   Figure 2

So, 342 × 657 = 224,694.

Use lattice grids to find the product.

1. 531 × 624 = _____   2. 248 × 715 = _____

Divisibility Rules!

Use the digits in the circles to form numbers that are divisible by the given divisors.

1.

1
2   5

Numbers divisible by 2: _____

Numbers divisible by 5: _____

2.

0
3   9

Numbers divisible by 3: _____

_____

Numbers divisible by 5: _____

3.

3
7   8

Numbers divisible by 2: _____

Numbers divisible by 9: _____

_____

4.

2
4   6

Numbers divisible by 4: _____

Numbers divisible by 6: _____

_____

5.

0
3   6

Numbers divisible by 3: _____

_____

Numbers divisible by 4: _____

Math Tip

Write the letter of the correct quotient from Column 2. Then, in the box at the bottom of the page, write the letter above the matching problem number. You will discover the math tip!

	Column 1	Column 2
_____	**1.** 351 ÷ 3	**A.** 70
_____	**2.** 747 ÷ 9	**B.** 29
_____	**3.** 156 ÷ 6	**C.** 63
_____	**4.** 210 ÷ 3	**D.** 117
_____	**5.** 256 ÷ 8	**E.** 67
_____	**6.** 133 ÷ 7	**I.** 83
_____	**7.** 74 ÷ 2	**L.** 47
_____	**8.** 116 ÷ 4	**M.** 22
_____	**9.** 448 ÷ 8	**N.** 56
_____	**10.** 315 ÷ 5	**O.** 32
_____	**11.** 486 ÷ 6	**P.** 97
_____	**12.** 188 ÷ 4	**R.** 19
_____	**13.** 477 ÷ 9	**S.** 53
_____	**14.** 194 ÷ 2	**T.** 26
_____	**15.** 201 ÷ 3	**U.** 81
_____	**16.** 176 ÷ 8	**V.** 78
_____	**17.** 468 ÷ 6	**Y.** 37

___	___	___	___	___	___	___	___	___	___
10	5	16	14	4	3	2	8	12	15

___	___	___	___	___	___	___		___	___	___	___	___	___
9	11	16	8	15	6	13		1	2	17	2	1	15

___	___	___	___	___	___
15	17	15	9	12	7

How High Can You Climb?

Can you get to the top of these steps? Solve the division problem at the bottom. Then put that answer on the next step and solve it. Continue to the top.

_____ ÷ _____ = 9

_____ ÷ 3 = _____

1. 540 ÷ 5 = _____

_____ ÷ _____ = 3

_____ ÷ 5 = _____

2. 960 ÷ 8 = _____

_____ ÷ _____ = 8

_____ ÷ 6 = _____

3. 720 ÷ 3 = _____

_____ ÷ _____ = 11

_____ ÷ 4 = _____

4. 616 ÷ 2 = _____

_____ ÷ _____ = 1

_____ ÷ 5 = _____

5. 860 ÷ 4 = _____

Name _____

Create a Problem

Write a word problem that could be solved with each division sentence given. Then solve your creation!

1. $237 \div 4 =$ _____

Problem _____

_____

_____

_____

_____

2. $637 \div 6 =$ _____

Problem _____

_____

_____

_____

_____

3. $822 \div 8 =$ _____

Problem _____

_____

_____

_____

_____

4. $207 \div 9 =$ _____

Problem _____

_____

_____

_____

_____

5. $735 \div 3 =$ _____

Problem _____

_____

_____

_____

_____

6. $517 \div 2 =$ _____

Problem _____

_____

_____

_____

_____

Number Patterns

Find the next three terms. Then write the rule used to form the pattern.

1. 26, 33, 47, 68, _____, _____, _____

   Rule: _____

2. 7.6, 9.1, 11.6, 15.1, _____, _____, _____

   Rule: _____

3. 631, 620, 610, 601, _____, _____, _____

   Rule: _____

4. 87, 91, 99, 111, 127 _____, _____, _____

   Rule: _____

5. 396, 391, 381, 366, _____, _____, _____

   Rule: _____

6. 17, 68, 34, 136, 68, _____, _____, _____

   Rule: _____

7. 23, 46, 138, 552, _____, _____, _____

   Rule: _____

Try making your own pattern. See if a classmate can figure it out.

Name _____

Patterns in Geometry

Draw the next three figures. Write the rule used to form the pattern.

1.  ☐ ◯ ◯ ☐ ☐   ___ ___ ___

Rule: _____

_____

2. 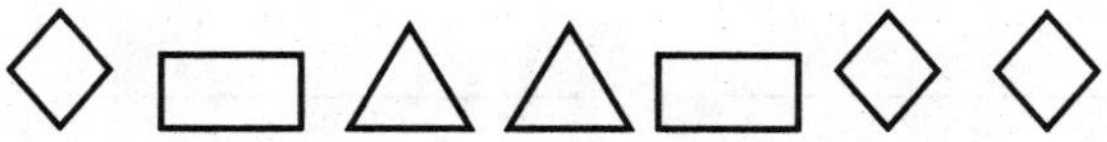 ◇ ▭ △ △ ▭ ◇ ◇

Rule: _____

_____

3. 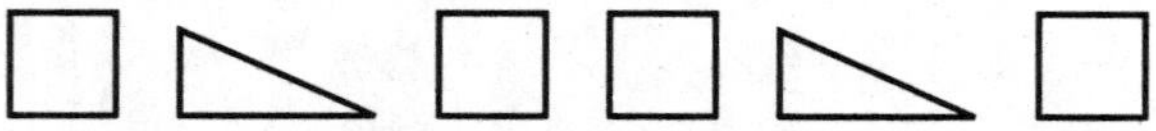 ☐ ◺ ☐ ☐ ◺ ☐

Rule: _____

_____

4. 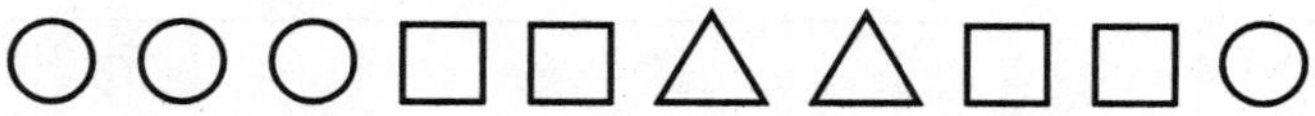 ◯ ◯ ◯ ☐ ☐ △ △ ☐ ☐ ◯

Rule: _____

_____

5.  ▽ ◇ ▽ ▽ ◇ ◇ ▽

Rule: _____

_____

Division Puzzles

Complete the number
sentences. Then solve
the puzzle.

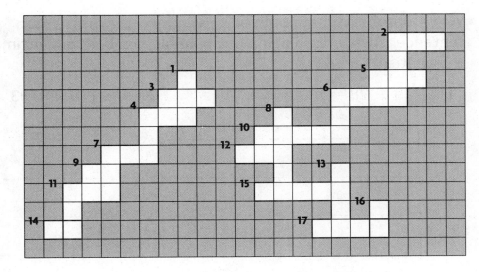

Across

1.   150 ÷ 50 = _____

2. 72,000 ÷ 80 = _____

3. 36,000 ÷ 90 = _____

4. 10,000 ÷ 20 = _____

5. 24,000 ÷ 60 = _____

6. 30,000 ÷ 6 = _____

7. 63,000 ÷ 90 = _____

8.   1,600 ÷ 200 = _____

9.   2,500 ÷ 50 = _____

10. 900,000 ÷ 30 = _____

11. 42,000 ÷ 70 = _____

12.   8,000 ÷ 80 = _____

13.   560 ÷ 70 = _____

14.   200 ÷ 20 = _____

15. 100,000 ÷ 5 = _____

16.   140 ÷ 20 = _____

17. 12,000 ÷ 6 = _____

Down

1.   9,000 ÷ 30 = _____

2. 810,000 ÷ 90 = _____

3.   800 ÷ 20 = _____

4. 35,000 ÷ 70 = _____

5.   1,600 ÷ 40 = _____

6. 45,000 ÷ 90 = _____

7. 42,000 ÷ 60 = _____

8. 400,000 ÷ 5 = _____

9.   3,000 ÷ 60 = _____

10.   180 ÷ 6 = _____

11. 18,000 ÷ 30 = _____

13. 560,000 ÷ 70 = _____

16.   4,900 ÷ 70 = _____

What's the Problem?

Write a problem that could be solved by using the division sentence. Then write a pair of compatible numbers, and estimate the quotient.

1. $1,489 \div 28 = n$

Problem: _____

_____

_____

_____

Compatible numbers:

_____

2. $7,100 \div 93 = n$

Problem: _____

_____

_____

_____

Compatible numbers:

_____

3. $63,147 \div 879 = n$

Problem: _____

_____

_____

_____

Compatible numbers:

_____

4. $276 \div 37 = n$

Problem: _____

_____

_____

_____

Compatible numbers:

_____

5. $758 \div 42 = n$

Problem: _____

_____

_____

_____

Compatible numbers:

_____

6. $41,797 \div 561 = n$

Problem: _____

_____

_____

_____

Compatible numbers:

_____

Number Pyramids

The numbers in the pyramids are found by using one of these simple formulas:

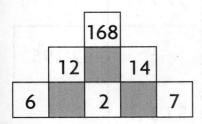

$A \times B = C$ **or**
$C \div A = B$ **or**
$C \div B = A$

If you know some of the numbers, you can find the ones that are missing.

To find the top number, multiply $12 \times 14 = 168$.

To find the number on the lower right, divide $14 \div 2 = 7$.

Fill in the missing numbers.

1.

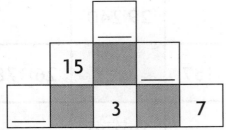

2.

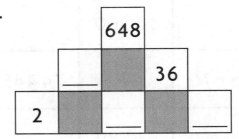

3.

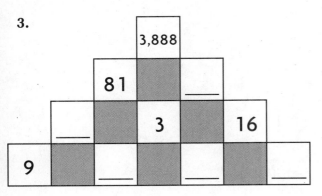

4.

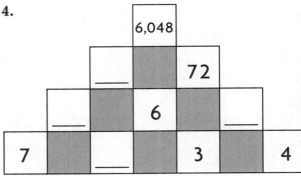

Division Crosses

The number in the center of each cross is an estimate
of a quotient. Look carefully at each division problem.

If the estimate is just right, color the problem red.
If the estimate is too low, color the problem green.
If the estimate is too high, color the problem blue.

1.

	83)260	
37)190	4	23)96
	61)250	

2.

	37)241	
19)101	6	42)295
	51)315	

3.

	36)300	
95)670	7	74)445
	83)589	

4.

	29)240	
16)157	8	26)178
	38)320	

Division Tip

Write the letter of the correct quotient from Column 2 on the
line in Column 1. Then write the letter on the line or lines with the
matching problem number in the box at the bottom of the page.

Column 1	Column 2
_____ 1. $842 \div 41$	A. 14 r3
_____ 2. $617 \div 53$	B. 5 r36
_____ 3. $119 \div 16$	C. 6 r31
_____ 4. $367 \div 26$	D. 11 r34
_____ 5. $509 \div 13$	E. 80 r41
_____ 6. $758 \div 65$	H. 8 r7
_____ 7. $496 \div 92$	I. 39 r2
_____ 8. $279 \div 34$	M. 62
_____ 9. $613 \div 72$	N. 51 r5
_____ 10. $547 \div 86$	O. 7 r7
_____ 11. $1,932 \div 35$	P. 59 r6
_____ 12. $6,014 \div 97$	R. 8 r37
_____ 13. $3,721 \div 46$	S. 20 r22
_____ 14. $5,745 \div 71$	T. 80 r65
_____ 15. $1,178 \div 23$	V. 55 r7
_____ 16. $2,602 \div 44$	W. 11 r43

___ ___ ___ ___ ___ ___ ___ ___   ___ ___   ___ ___ ___ ___ ___ ___ ___   ___ ___ ___
9 13 12 13 12 7 13 9   14 3   10 3 12 16 4 9 13   14 8 13

___ ___ ___ ___ ___ ___ ___ ___ ___   ___ ___ ___ ___   ___ ___ ___   ___ ___ ___ ___ ___ ___ ___
9 13 12 4 5 15 2 13 9   6 5 14 8   14 8 13   2 5 11 5 1 3 9

___ ___   ___ ___ ___ ___   ___ ___ ___ ___ .
4 14   13 4 10 8   1 14 13 16

Operation 36

There are 4 numbers in each circle. Use each number once to perform 3 different operations and write the number sentences. Use the answer for A in your number sentence in B, and your answer for B in your number sentence in C. Your answer for C should be 36. The first one has been done for you.

1.

(2)
(15   3)
(12)

A. _____ $2 \times 12 = 24$ _____

B. _____ $24 + 15 = 39$ _____

C. _____ $39 - 3 = 36$ _____

2.

(2)
(4   7)
(22)

A. _____

B. _____

C. _____

3.

(0)
(2   3)
(54)

A. _____

B. _____

C. _____

4.

(4)
(5   16)
(100)

A. _____

B. _____

C. _____

5.

(4)
(8   9)
(18)

A. _____

B. _____

C. _____

6.

(2)
(8   3)
(10)

A. _____

B. _____

C. _____

Number Patterns

Find the next three terms.
Write the rule used to form the pattern.

1. 24, 30, 25, 31, 26, _____, _____, _____

Rule: _____

2. 3, 24, 12, 96, 48, _____, _____, _____

Rule: _____

3. 27, 32, 42, 57, 77, _____, _____, _____

Rule: _____

4. 500, 100, 1,000, 200, 2,000, _____, _____, _____

Rule: _____

5. 37, 222, 74, 444, _____, _____, _____

Rule: _____

6. 19, 56, 103, 160, _____, _____, _____

Rule: _____

7. 635, 610, 575, 530, 475, _____, _____, _____

Rule: _____

8. 42, 21, 126, 63, 378, _____, _____, _____

Rule: _____

9. 579, 572, 582, 575, 585, _____, _____, _____

Rule: _____

10. 18, 162, 54, 486, 162, _____, _____, _____

Rule: _____

More on Median and Mode

Sometimes there is not a single number in the middle of a set of numbers. For example, look at the data set at the right. The numbers 7 and 9 are both in the middle.

Data Set: 5, 7, 9, 10

To find the median of a group of numbers that shows no one number exactly in the middle, add the two middle numbers and divide by 2.

$7 + 9 = 16$
$16 \div 2 = 8$
The median is 8.

The mode is the number that occurs most often. There can be more than one mode or no mode. Look at the examples below.

a. 2, 3, 4, 5

No mode

b. 2, 2, 3, 4,

The mode is 2.

c. 2, 2, 3, 3, 4

There are two modes: 2 and 3.

d. 2, 2, 3, 3, 3, 4, 4

The mode is 3.

Order numbers in each set from least to greatest. Then find the median. You may use a calculator.

1. 7, 9, 6, 4, 3, 2

2. 12, 5, 9, 1

3. 24, 32, 20, 30

_____

_____

_____

4. 1, 7, 5, 6

5. 23, 46, 78, 21, 35, 98

6. 82, 77, 64, 65, 63, 60

_____

_____

_____

Find the mode or modes for each set of numbers.

7. 4, 5, 6, 7, 5

8. 12, 45, 16, 11

_____

_____

9. 1, 5, 1, 6, 5, 6, 9

10. 5, 8, 2, 5, 8, 1, 2, 5

_____

_____

Pick One!

Once there were three rats called Mean, Median, and Mode. They were given their names because the numbers they ate were best described by the mean, median, or mode of the numbers.

Find the mean, median, and mode for each rat. Decide if the mean number, median number, or mode number best describes most of the numbers inside the rat. Then name each rat Mean, Median, or Mode.

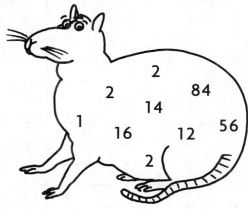

_____

Name: _____

_____   _____

Name: _____   Name: _____

Selective Scales

Najuma wants to display the number of phone calls she gets each day for a week.

Phone Calls				
Mon	**Tue**	**Wed**	**Thu**	**Fri**
2	10	8	7	24

Use the data in the table to complete each graph. Use the intervals noted below each graph.

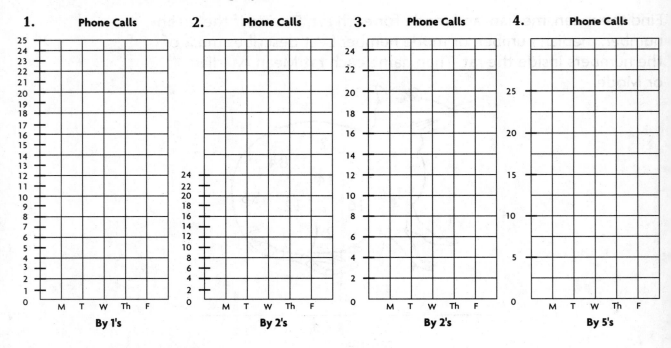

1. Phone Calls — By 1's

2. Phone Calls — By 2's

3. Phone Calls — By 2's

4. Phone Calls — By 5's

For Problems 5–7, use the graphs above.

5. Which graph uses an interval that makes the best use of the graph space?

   Explain. _____

   _____

6. Which graph was the hardest to draw data points on?

   Explain. _____

   _____

7. What is similar and what is different about graphs 2 and 3?

   _____

   _____

Breakfast Business

SALES COMPARISON

Sales in Millions ($)

Year

——————— **Better Bagels**   - - - - - - - - - - **California Cocoa**

Use the Sales Comparison graph to compare each business.

1. In 1980, how much money did each company make?

_____

2. In 1986, how much money did each company make?

_____

3. In 1990, how much money did each company make?

_____

4. In what years labeled on the graph did Better Bagels'
   profits exceed California Cocoa's profits?

_____

5. In 1990, one of the companies opened up new stores that
   helped to increase sales by 30 million dollars after 2 years.
   Which company was this?

_____

6. Write a letter to the president of one of the companies.
   Explain, based on the company's profits, why you would
   like to work for that particular company.

Which Is Which?

Graphs A and B were made using the same data.

Graph A

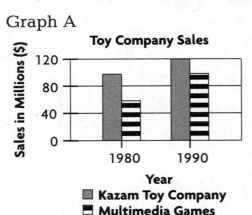

Graph B

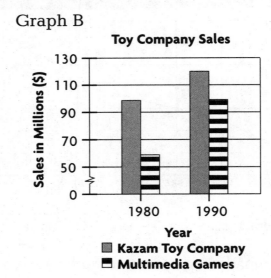

1. You are a member of the Kazam Toy Company. Which graph would you use to convince others how well your company is doing compared to the competition. Explain your reasoning.

_____

_____

2. You are a member of Multimedia Games. Which graph would you use to convince others how well your company is doing compared to the competition. Explain your reasoning.

_____

_____

3. Why do the two graphs look so different? Explain.

_____

_____

4. What interval would more honestly represent both companies' sales?

_____

_____

Home on the Range

Ranger Kate keeps track of antelope herd size each month for a year.

This year the range in the number of antelopes in the herd is $86 - 23 = 63$.

Herd Size
Jan -35   Jul-68
Feb-42   Aug-86
Mar-60   Sept-75
Apr-55   Oct-64
May-72   Nov-52
Jun -74   Dec-23

23   63   86

Find the range for each of these other herd counts:

1. _____

Herd Size
Jan-35   Jul-68
Feb-42   Aug-36
Mar-50   Sept-75
Apr-55   Oct-84
May-72   Nov-52
Jun-64   Dec-32

2. _____

Herd Size
Jan-125   Jul-141
Feb-106   Aug-192
Mar-138   Sept-167
Apr-111   Oct-173
May-148   Nov-145
Jun-156   Dec-132

3. _____

Herd Size
Jan-103   Jul-122
Feb-97   Aug-93
Mar-111   Sept-87
Apr-114   Oct-98
May-125   Nov-96
Jun-137   Dec-91

4. _____

Herd Size
Jan-214   Jul-235
Feb-185   Aug-212
Mar-176   Sept-228
Apr-191   Oct-212
May-211   Nov-195
Jun-223   Dec-200

Slice That Pie!

Answer the questions about each graph.

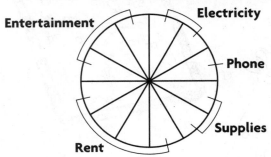

**JOE'S COMPUTER SERVICE
$2,400 MONTHLY OFFICE BUDGET**

Entertainment · Electricity · Phone · Supplies · Rent

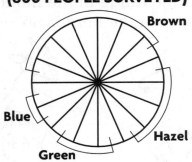

**EYE COLOR
(800 PEOPLE SURVEYED)**

Brown · Hazel · Green · Blue

1. How much money is in Joe's monthly office budget?

_____

2. How many parts is the circle graph cut into?

_____

3. How much money does each part of the graph represent?

_____

4. About how much money does Joe pay for entertainment? electricity? phone? supplies? rent?

_____

_____

Make a circle graph.

5. Joe has cut his entertainment expenses to $400. Make a circle graph to show what Joe's new budget will look like.

6. How many people were surveyed?

_____

7. How many parts is the circle graph cut into?

_____

8. How many people does each part of the graph represent?

_____

9. About how many people had brown eyes? blue eyes? green eyes? hazel eyes?

_____

_____

Make a circle graph.

10. Use the data from the graph above to make a graph that is divided into 8 parts. Let each part represent 100 people.

Garden Graphs

The Green Ivy School plants several gardens each spring.
Complete each table. Use fraction-circle pieces to make a circle
graph that shows how each garden is planted.

1.

BEAN GARDEN			
Bean	Number of Rows	Fraction	Simplified Fraction
Lima	15	$\frac{15}{30}$	$\frac{1}{2}$
Snap	10	$\frac{10}{30}$	$\frac{1}{3}$
Fava	5	$\frac{5}{30}$	$\frac{1}{6}$

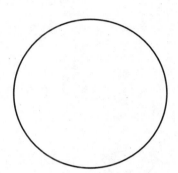

2.

HERB GARDEN			
Herb	Number of Packages	Fraction	Simplified Fraction
Rosemary	2		
Oregano	2		
Cumin	4		
Chives	4		
Parsley	4		

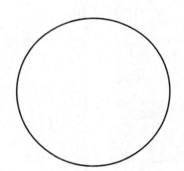

3.

POTATO GARDEN			
Potato	Number of Plants	Fraction	Simplified Fraction
Yukon Gold	15		
Frontier	15		
Acadia Russet	15		

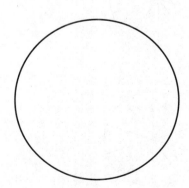

Circle Logic

You can make 8 whole circle graphs using the decimals below.
Use each decimal only once.

0.1, 0.1, 0.1, 0.1, 0.1, 0.1,

0.2, 0.2, 0.2, 0.2, 0.2, 0.2,

0.3, 0.3, 0.3, 0.3, 0.3,

0.4, 0.4, 0.4,

0.5, 0.6, 0.7, 0.8, 0.9

1.

2.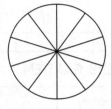

3.

4.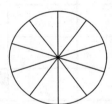

5.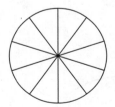

6.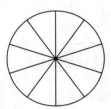

7.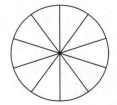

8.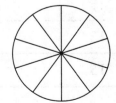

Find the Error!

Mike would like to use a double-line graph to display the data in the table.

4-MONTH SALES RECORD		
Months	**Company A**	**Company B**
January	$120 million	$140 million
February	$160 million	$180 million
March	$100 million	$100 million
April	$140 million	$120 million

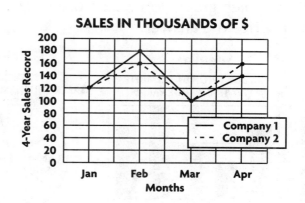

SALES IN THOUSANDS OF $

There are 8 errors in Mike's double-line graph above.
Find each error and explain how each should be fixed.
Then draw an accurate double-line graph using the data table above.

1. _____

2. _____

3. _____

4. _____

5. _____

6. _____

7. _____

8. _____

Tracking Tourney Trivia

Every year Tremont Middle School hosts a 5-school track
meet. Each graph below represents different information.
Write which graph you would use to answer each question.
Then write the answer.

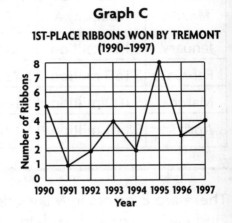

Graph A

RIBBONS WON BY TREMONT
IN 1997 TRACK TOURNAMENT

Graph B

1997 TREMONT
TRACK TOURNAMENT BUDGET

Graph C

1ST-PLACE RIBBONS WON BY TREMONT
(1990–1997)

1. What is Tremont's budget for the track tournament?

_____

2. In what year did Tremont win the most 1st-place ribbons?

_____

3. How many 1st-place ribbons did Tremont win in 1997?

_____

4. How many place ribbons did Tremont earn in 1997?

_____

5. How many 1st-place ribbons did Tremont win during the tournaments?

_____

6. The Tremont Tournament has 5 events and awards ribbons for the first
   5 places. Every place ribbon costs the same amount to purchase. How
   much money did Tremont School spend on ribbons that were earned
   by their own students?

_____

Dollar Data

The U.S. Treasury Department is responsible for making sure there are enough of each type of bill in circulation.

1. Complete the table.

NUMBER OF U.S. BILLS IN CIRCULATION IN 1994		
Type of Bill	**Amount in Circulation**	**Number of Bills in Circulation**
$1 bills	$5.6 trillion	5.6 trillion
$2 bills*	$0.8 trillion	0.4 trillion
$5 bills	_____ trillion	1.3 trillion
$10 bills	_____ trillion	1.3 trillion
$20 bills	_____ trillion	3.7 trillion
$50 bills	_____ trillion	0.8 trillion
$100 bills	_____ trillion	2.0 trillion

*Not printed since 1977.

2. Complete the bar graph with the data from the table above.

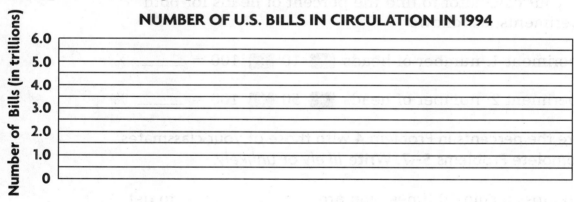

NUMBER OF U.S. BILLS IN CIRCULATION IN 1994

Number of Bills (in trillions)

6.0
5.0
4.0
3.0
2.0
1.0
0

3. About how many trillion dollars were in circulation?

_____

4. Compare the number of $10 and $20 bills in circulation.

_____

Don't Flip Out!

A coin will land heads half of the time.

What if you toss a coin 10 times? Are you likely to get 5 heads and 5 tails?

What if you toss a coin 50 times? Are you likely to get 25 heads and 25 tails?

Try these experiments before you answer.

Heads	Tails	Total
		10
		50

1. Toss a coin 10 times. Record your tallies in the table.

2. Toss a coin 50 times. Record your tallies in the table.

3. Compare your results with those of your classmates. How many students got exactly 5 heads and 5 tails? How many students got exactly 25 heads and 25 tails?

   _____

4. Use your calculator to find the percent of heads for both experiments, as follows.

   Experiment 1: number of heads ➗ 10 ✖ 100 = _____ %

   Experiment 2: number of heads ➗ 50 ✖ 100 = _____ %

Compare the percents in Problem 4 with those of your classmates. Then complete Problems 5–7. Write *likely* or *unlikely*.

5. If you toss a coin 10 times, you are _____ to get exactly 10 heads.

6. If you toss a coin 50 times, you are _____ to get exactly 50 heads.

7. If you toss a coin 50 times, you are _____ to get between 40% and 60% heads.

Summing It Up

What happens when you roll two cubes, each numbered 1–6?

To find out, answer the following questions.

1. Make a list of all possible outcomes of rolling the two number cubes. 1 and 1; 1 and 2; 1 and 3;

_____

_____

_____

_____

2. List the sums of each possible outcome in problem 1. For example, the sum for the outcome 1 and 4 is 5.

_____ 2, 3, 4, _____

3. Which sums are most likely to occur?

_____

4. Which sums are least likely to occur?

_____

5. Predict the number of times each sum will occur if you roll the number cubes 50 times. Then test your prediction by rolling the cubes 50 times. Use tally marks to record the results in the table.

Sum	2	3	4	5
Predicted frequency				
Actual frequency				

6. Did your actual results match your predicted outcomes? Explain.

_____

_____

_____

The Path of Probability

Toss a coin 5 times to follow a probability path from the start to the end boxes.

Rules 1. Toss the coin. If it is heads, follow the heads path to the next oval. If it is tails, follow the tails path.

  2. Put a tally mark in an oval for each toss.

  3. After 5 tosses, record the letter of the box in which you land.

  4. Repeat the process 20 times.

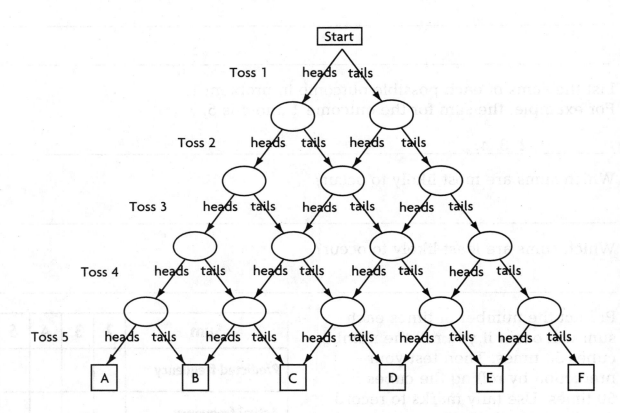

1. In which boxes did you land most often?

_____

2. In which boxes did you land least often?

_____

Alphabet Probability

1. Predict the 5 most commonly used letters in the English

   language. _____

2. Pick 5 or more sentences from a book or a magazine. In the table below, tally how many times each letter occurs.

Letter	A	B	C	D	E	F	G	H	I	J	K	L	M
Tally													

Letter	N	O	P	Q	R	S	T	U	V	W	X	Y	Z
Tally													

3. List your results in order from greatest to least frequency.

   _____

Here is an authentic list of letters in the English language in order from the greatest to the least frequency of use.

E, T, A, O, N and I, S, R, H, L, D, C, U, P and F, M, W, Y, B and G, V, K, Q and X, J and Z

Those letters with an *and* have the same frequency of occurrence.

4. Why is your list not the same as the authentic list?

   _____

   _____

5. What do you notice about these three sentences?
   This is odd. Do you know why? Try and find out.

   _____

   _____

Three Coins in a Fountain

When you toss a coin, there are just two possible outcomes: heads or tails.

If you toss two coins at once, there are three possible outcomes:

- 2 heads
- 1 head and 1 tail
- 2 tails

For Problems 1–2, complete the sentence.

1. If you toss three coins at once, there are four possible outcomes: 3 heads, 2 heads and 1 tail,

   _____ and _____.

2. If you toss four coins at once, how many possible outcomes are there? What are they?

   _____

   _____

For Problems 3–4, use the table.

Try this experiment. Toss two coins at once, and tally the results of the tosses. Repeat for a total of 20 tosses.

2 Heads	1 Head and 1 Tail	2 Tails

3. Of the 20 tosses, how many times did you get 2 heads? 1 head and 1 tail? 2 tails?

   _____

4. Compare your results with those of your classmates. Which outcome seems more likely: 2 tails or 1 head and 1 tail?

   _____

Presidential Probability

Does the president's first name help him to get elected?

Certain first initials, such as *J*, *W*, and *G*, seem to suggest an answer to this question. Of the first 42 presidents of the United States, half of them had first names beginning with one of these letters. Here is the data.

First Letter of U.S. Presidents' First Names												
A	**B**	**C**	**D**	**E**	**F**	**G**	**H**	**I**	**J**	**K**	**L**	**M**
3	1	2	1	0	2	5	2	0	10	0	1	2
N	**O**	**P**	**Q**	**R**	**S**	**T**	**U**	**V**	**W**	**X**	**Y**	**Z**
0	0	0	0	3	0	2	1	0	6	0	0	1

1. List the top 10 letters for the presidents' first initial in order from greatest to least.

_____

2. You pick a president at random among the first 42 presidents. What is the probability that his name begins with *J*? *W*? *E*?

_____

3. Why do you think so many presidents have first initial *J*?

_____

4. Two candidates are running for the office of president. One has the first name Barbara, and the other Geraldine. Explain whether you think one has an advantage based on her first initial.

_____

_____

_____

Name _____

How Much Money?

There is $0.45 in change in a sealed envelope. How good are your chances of winning the contest? What are all the possible coin combinations?

1. There are at least 31 possible combinations. Make an organized list to find all the possible coin combinations.

	Penny	Nickel	Dime	Quarter
1)	45	0	0	0
2)	40	1	0	0
3)	35	0	1	0
4)	35	2	0	0
5)	30	1	1	0
6)	30	3	0	0
7)	25	0	2	0
8)				
9)				
10)				
11)				
12)				
13)				
14)				
15)				
16)				

	Penny	Nickel	Dime	Quarter
17)				
18)				
19)				
20)				
21)				
22)				
23)				
24)				
25)				
26)				
27)				
28)				
29)				
30)				
31)				

2. Suppose you are the only person entering the contest. What are your chances of winning?

3. What are your chances of winning if you get a new clue that there are fewer than 20 coins in the envelope?

4. What are your chances of winning if you find that there are only 3 coins in the envelope?

_____ _____ _____

Name _____

Tiny Numbers

What comes after the ten-thousandths place? Use the pattern to
complete the table.

$2 \times 0.1 = 0.2$	two $\times$ one tenth = two tenths
$2 \times 0.01 = 0.02$	two $\times$ one hundredth = two hundredths
$2 \times 0.001 = 0.002$	two $\times$ one thousandth = two thousandths
1. $2 \times 0.0001 = $ _____	two $\times$ one ten-thousandth = two ten-thousandths
2. $2 \times 0.00001 = $ _____	two $\times$ one hundred-thousandth = two hundred-thousandths
3. $2 \times 0.000001 = $ _____	two $\times$ one millionth = two millionths
4. $2 \times 0.0000001 = $ _____	two $\times$ one ten-millionth = two ten-millionths
5. $2 \times 0.00000001 = $ _____	two $\times$ one hundred-millionth = two hundred-millionths

Complete the next 2 rows of the table.

6. $2 \times$ _____ $= 0.000000002$	two $\times$ _____ = two billionths
7. $2 \times 0.0000000001 = $ _____	two $\times$ one ten-billionth = _____

Write the place-value of the product if you multiplied:

8. tenths $\times$ tenths

_____

9. hundredths $\times$ tenths

_____

10. tenths $\times$ ten-thousandths

_____

11. tenths $\times$ millionths

_____

12. hundredths $\times$ hundred-thousandths

_____

13. thousandths $\times$ thousandths

_____

Mystery Triangles

The numbers in the squares are the products of the numbers in the circles. Find the missing numbers.

1.

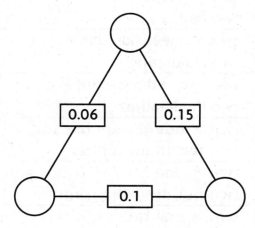

2.

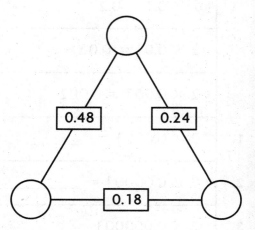

3.

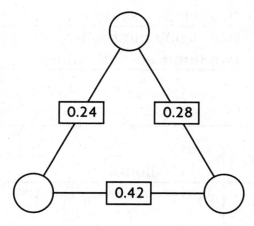

4.

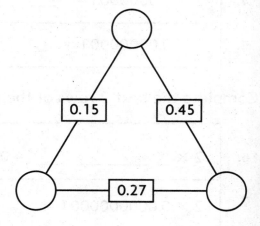

5.

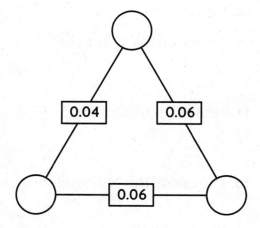

6.

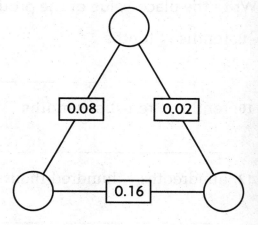

Circle Patterns

1. What is the greatest number of pieces you can cut the circle into by using only one line?

_____

2. What is the greatest number of pieces you can cut the circle into by using two lines?

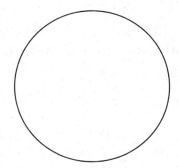

_____

3. What is the greatest number of pieces you can cut the circle into by using three lines?

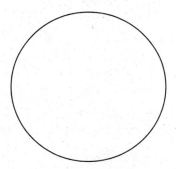

_____

4. What is the greatest number of pieces you can cut the circle into by using four lines?

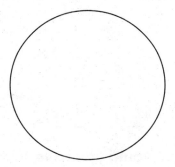

_____

5. Make a list of the number of pieces you made in circles 1– 4. What pattern do you see in the increasing number of pieces?

_____

_____

_____

6. Suppose you were to make 5, 6, and 7 cuts in circles. Use the pattern to predict the greatest number of pieces you can make in each circle.

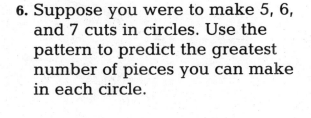

_____

_____

_____

Money Grows on Trees!

A banker has promised to help you make money. For every month you leave money with him, the amount doubles.

You give him $1.23. He puts it on a branch of his money tree, and soon another branch grows another $1.23.

You now have 2 × $1.23 = $2.46.

One month later ...

When you leave the money for one more month, four branches grow with $1.23 on them.

You now have 2 × $2.46 = $4.92.

Two months later ...

How much would you have if you invested each amount of money for 2 months?

1.  $0.85
_____

2. $2.94
_____

3.  $3.58
_____

4. $6.37
_____

How much would you have if you invested each amount of money for 6 months?

5. $0.85
_____

6. $2.94
_____

7.  $3.58
_____

8. $6.37
_____

9. $4.79
_____

10. $19.35
_____

Shopping Madness

You get to do your own shopping for school. You are given $50.00 to buy as many supplies as you can.

Use the price list at the right to see how you can spend as close to $50.00 as possible, without going over. You must buy at least one of each item.

SCHOOL SUPPLIES	
Item	Cost
Pencil	$0.35
Pen	$0.64
Folder	$0.89
Notebook	$1.40
Binder	$4.48
Backpack	$22.52

Fill in the chart.

Pencils	Pens	Folders	Notebooks	Binder	Backpack	Total
10	10	10	3	1	1	$50.00
$3.50	$6.40	$8.90	$4.20	$4.48	$22.52	

Missing Numbers

Find the missing numbers in the multiplication problems. Put the decimal point where it belongs in the final answer.

1.

```
        8. □ □
    ×   0. □ 6
    ───────────
        5 □ 4 4
    □   6 2 □ 0
    ───────────
    3 □ 4 6 4
```

2.

```
      □ .7 □
    ×  0 .□ 6
    ───────────
        □ 2 6
      2 1 3 □
    ───────────
    0 □ 5 5 6
```

3.

```
        2 .9 □
    ×      □ .7
    ───────────
      2 □ 9 3
    1 □ 9 □ 0
    ───────────
    □ 0 0 3 3
```

4.

```
      1 □ .3
    ×    4. □
    ───────────
      □ 8 4
    □ 9 □ 0
    ───────────
    □ 9 0 4
```

5.

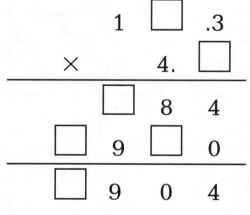

```
      □ □ . □
    ×  □ . 5 4
    ───────────
      1 4 5 □
    1 □ 2 □ □
    3 6 4 □ □
    ───────────
    □ 6 □ 5 6
```

6.

```
        □ 5 .6
    ×   □ . 1 □
    ─────────────
      □ 9 0 4
      6 □ 6 0
    1 3 1 2 □ 0
    ─────────────
    □ 4 □ 6 □ 4
```

Number Maze

Solve.

$300 \div 6 =$ _____ $30 \div 6 =$ _____ $3 \div 6 =$ _____

$200 \div 5 =$ _____ $20 \div 5 =$ _____ $2 \div 5 =$ _____

$400 \div 8 =$ _____ $40 \div 8 =$ _____ $4 \div 8 =$ _____

$800 \div 8 =$ _____ $80 \div 8 =$ _____ $8 \div 8 =$ _____

$100 \div 4 =$ _____ $10 \div 4 =$ _____ $1 \div 4 =$ _____

Find the division sentence in the maze. Problems can be written horizontally, vertically, and diagonally.

4	3	0	0.4	2	0	5	4	4	8	0	1
0	0	3	2	5	8	0	0	8	0.1	0.4	5
0	0	2	3	0.4	1	2.5	0	0	4	2	5
8	6	0	0	0	0	1	8	8	1	2	0.4
5	5	0	6	4	4	3	5	1	4	0.25	0
2	0	1	5	0	2.5	6	0	0	3	2	0
3	0	0	6	0.5	0	0.5	0	1	0	0	3
0	0	0.5	0	0	4	1	0	0.25	8	4	0.4
2	0	0	4	0	8	5	5	0	0	2.5	0
1	3	0	0	0	0.5	4	0	0.5	2	0	0
4	0	8	0	4	0	1	0	0.25	0	0	2
0.5	0	8	0	0	8	1	0	8	8	1	5

Find the Fuzzles

What does a 0.04 fuzzle look like?

Here are some 0.04 fuzzles. They can look like any of the following pieces.

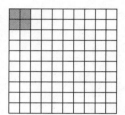 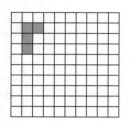 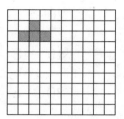 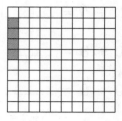

These pieces are not fuzzles.

 or

Fuzzle pieces must all be connected by a side.

The hundredths square below can be divided into 9 different-size fuzzle pieces.

Find each of these fuzzles in the hundredths square below. Outline each fuzzle you find, and label it with its decimal equivalent.

0.02, 0.03, 0.04, 0.07, 0.08, 0.09, 0.17, 0.20, and 0.30

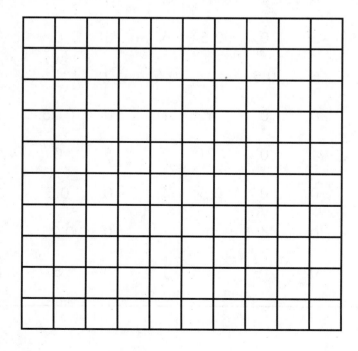

Share and Share Alike!

The Chocolate Lab Dog Bar makers want to advertise that the bar is big enough to share.

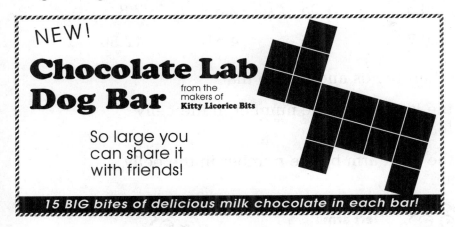

NEW!

Chocolate Lab
Dog Bar
from the makers of
Kitty Licorice Bits

So large you can share it with friends!

15 BIG bites of delicious milk chocolate in each bar!

There are many ways to split the bar in half. Here are two ways.

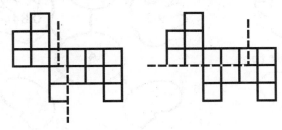

Each person will get 7.5 chunks.

Show two ways to split the bars. Tell how many chunks each person gets.

1. if 3 people share

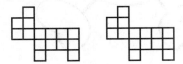

2. if 4 people share

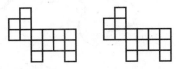

_____

3. if 5 people share

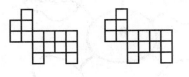

4. if 6 people share

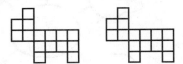

_____

Right to Left and Left to Right

Each person below is missing the decimals in their hands or feet. The missing numbers are

0.03	0.21	0.5	1.35	2.3	7.8
0.07	0.32	0.73	0.36	4.51	12.36

Place the numbers in their hands and feet so that

1. when you divide the left arm by the number in the belly you get the right foot.

2. when you divide the right arm by the number in the belly you get the left foot.

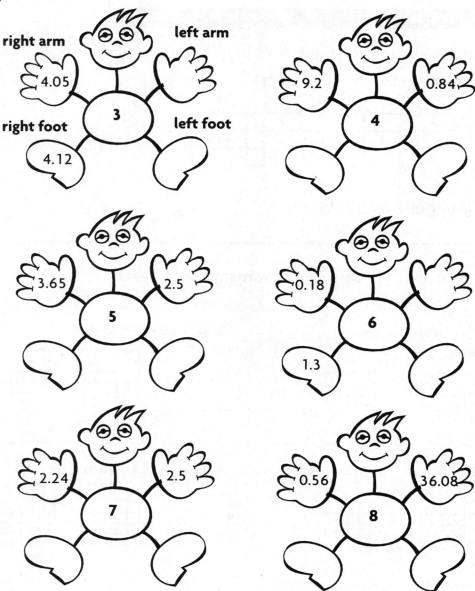

Number Puzzle

Find the sum of the quotients along each diagonal.

1.

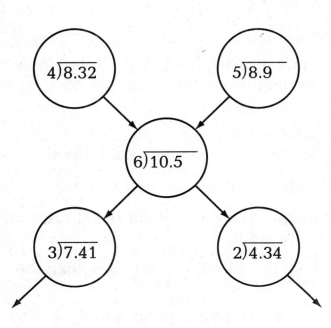

_____   _____

2.

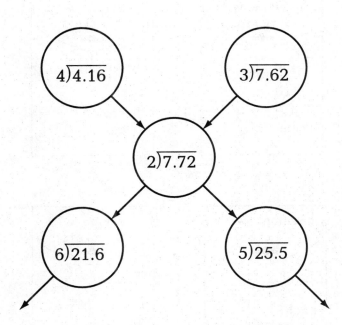

_____   _____

Tricky Tickets!

Eight groups of people bought tickets to different events. The envelopes containing the tickets got mixed up. Use the clues below to help match the envelopes with the amount of money each group paid.

Clues:

1. Each group's total amount paid was evenly divisible by the number of people in their group. For example, $9.45 ÷ 9 = $1.05 each.

2. Each group has a different number of people in it. The groups have 2, 3, 4, 5, 6, 7, 8, and 9 people in them.

3. Each group paid a different price for their tickets.

4. The group with 2 people in it got a better ticket price than the group with 3 people in it.

5. The group of 4 people got a better ticket price than the group of 6 people.

6. The prices paid were: $5.67, $7.14, $9.20, $10.92, $12.57, $13.75, $14.21, and $19.14.

Envelope with
2 tickets

Total: _____

Each
Ticket
Price: _____

Envelope with
3 tickets

Total: _____

Each
Ticket
Price: _____

Envelope with
4 tickets

Total: _____

Each
Ticket
Price: _____

Envelope with
5 tickets

Total: _____

Each
Ticket
Price: _____

Envelope with
6 tickets

Total: _____

Each
Ticket
Price: _____

Envelope with
7 tickets

Total: _____

Each
Ticket
Price: _____

Envelope with
8 tickets

Total: _____

Each
Ticket
Price: _____

Envelope with
9 tickets

Total: _____

Each
Ticket
Price: _____

Name _____

Can You Measure Up?

Estimate the length of the following objects. Then measure, using a centimeter ruler. Write both your estimate and exact measurement.

 **1.**    **2.**   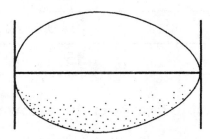 **3.**

Estimate _____   _____   _____

Actual   _____   _____   _____

4. 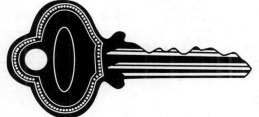   **5.**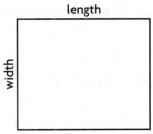

Estimate _____   _____

Actual   _____   _____

Measure the length and width for each figure in mm and cm.

6.
width / length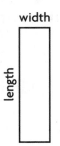

7. width / length

8. length / width

length _____   length _____   length _____

_____   _____   _____

width _____   width _____   width _____

_____   _____   _____

Fish Scales

Joe, Joey, and Joanne entered a fishing tournament. The fish they caught were measured using different types of metric scales.

There was a kilogram-only scale, a gram-only scale, and a milligram-only scale. Use your knowledge of the metric system to help judge who will win the fishing contest.

FISHING TOURNAMENT	
Contestant	**Weight of Fish Caught**
Joe	4,000 g; 3,500 g; 2,000,000 mg; 1,500,000 mg; 3 kg; 4.5 kg
Joey	3,000 g; 2,000 g; 1,500,000 mg; 4,000,000 mg; 2.5 kg; 1 kg
Joanne	4,500 g; 3,000 g; 2,500,000 mg; 5,000,000 mg; 3.5 kg; 2 kg

Use the table to answer the following questions.

1. List Joe's 3 largest fish from greatest to least weight.

_____

2. List Joey's 3 largest fish from greatest to least weight.

_____

3. List Joanne's 3 largest fish from greatest to least weight.

_____

4. Who caught the largest fish? What was its weight?

_____

5. List the weights of Joey's 6 fish in grams.

_____

_____

6. List the weights of Joanne's 6 fish in kilograms.

_____

How Full Is It?

Each of the containers below holds a certain amount when full. Shade each container to show about the number of mL or L shown below. The first container is shaded for you.

This vase holds 1 liter.

| 300 mL | 1L | 800 mL | 200 mL |

1. A large glass will hold 500 milliliters.

| 250 mL | 400 mL | 375 mL | 300 mL |

2. An olive oil bottle holds 750 milliliters.

| 187.5 mL | 750 mL | 375 mL | 250 mL |

3. A soda bottle holds 2 liters.

| 1L | 1.5 L | 500 mL | 1,000 mL |

Adding It Up

Match the sums in Column B with the equations in Column A.
You will need to change the units of some measures to compute
the answer.

	Column A	Column B
1. _____	4,000 m + 2 km	A. 110 cm
2. _____	6 dm + 20 cm + 300 mm	B. 22 m
3. _____	30,000 dm + 6 km	C. 408 cm
4. _____	48 dm + 36 cm + 22 mm	D. 6 km
5. _____	12,000 mm + 60 dm + 4 m	E. 4.5 km
6. _____	40 mm + 4 cm + 4 m	F. 9 km
7. _____	500,000 cm + 200 m	G. 172.5 cm
8. _____	600 mm + 2 dm + 30 cm	H. 11 dm
9. _____	6 km + 900 m + 10,000 cm	I. 4 km
10. _____	3,000,000 mm + 1,000 m	J. 5.2 km
11. _____	130 cm + 425 mm	K. 518.2 cm
12. _____	23 m + 38 dm + 22 cm	L. 20,700 m
13. _____	4 km + 180 m + 320 m	M. 2,702 cm
14. _____	20 km + 400 m + 300 m	N. 7 km

Metric Dominoes

Play this game with a partner.

Make four copies of each domino below.

To play: 1. Give each player six dominoes. Place the remaining dominoes in a pile.

2. Player 1 puts down a domino.

3. Player 2 puts down a domino with a metric equivalent. For example, 30 dm and 3 m are equivalents.

4. If a player does not have a metric equivalent, that player draws from the pile until a match is possible.

5. The first player to use all of his or her dominoes wins!

30 dm	3 km

0.03 m	30 cm

3 dm	3 cm

3 m	30,000 dm

30 mm	3,000 mm

3,000 m	0.3 m

Runner's Log

Sharon kept a log of her daily track workouts for one week.

Sunday	Ran 15 km
Monday	Ran 2,300 m in morning
	Interval workout in afternoon
	100 m — 7 times
	200 m — 6 times
	1,100-m run
Tuesday	Ran 3.85 km in morning
	Ran 10,000 m in afternoon
Wednesday	Ran 5.5 km in morning
	Interval workout in afternoon:
	400 m — 4 times
	200 m — 4 times
Thursday	Ran 6,000 m in morning
	Ran 8,000 m in afternoon
Friday	Ran 1,000 m in morning
	Interval workout in afternoon:
	15,000 m — 2 times
	800 m — 4 times
Saturday	Track meet:
	Ran 1,500 m and 5,000 m

Use Sharon's log to answer the following questions.

1. What is the total distance Sharon ran

   in kilometers? _____ in meters? _____

2. List the three days Sharon ran the farthest. Include the
   distances, from most to least.

   _____

   _____

3. If Sharon continues to work out at the same pace, how
   many weeks until she runs 1,000 km? _____

What Fraction Am I?

For each shape shown, tell what fraction of the shape is shaded.
Some of the shapes have two possible answers.

1.

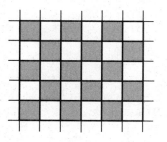

_____

2.

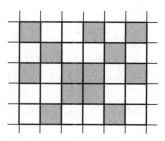

_____

3.

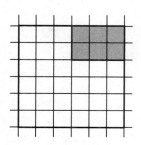

_____

4.

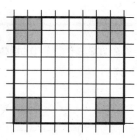

_____

5.

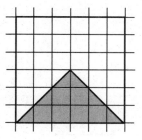

_____

6.

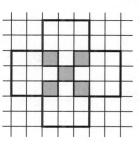

_____

7.

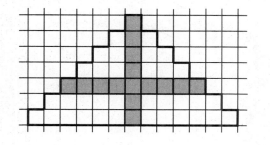

_____

8.

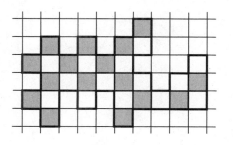

_____

Greater Than One!

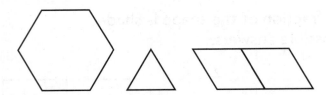

In figure A if the triangle represents the number 1, the hexagon represents the number 6.

If the hexagon represents 1, the triangle represents $\frac{1}{6}$.

In figure B if the parallelogram represents the number 1, the triangle represents the fraction $\frac{1}{2}$, and figure B represents $1\frac{1}{2}$.

For each figure and fraction, determine which shape represents the number 1.

1.

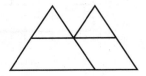

$2\frac{1}{3}$

_____

_____

2.

$10\frac{1}{2}$

_____

_____

3.

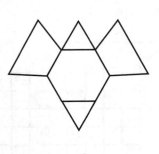

$2\frac{2}{6}$

_____

_____

4.

$9\frac{1}{3}$

_____

_____

What's Next?

Draw the next figure in each pattern.

1.
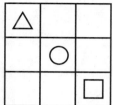 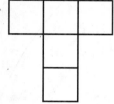 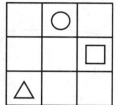

2.
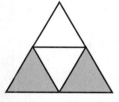

3.
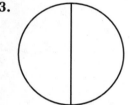

4.
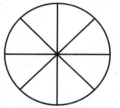

5.
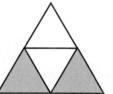

6.
 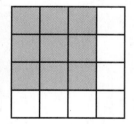 

Made in the Shade

Shade each section of the circle in which the fraction on the outside is greater than the one on the inside.

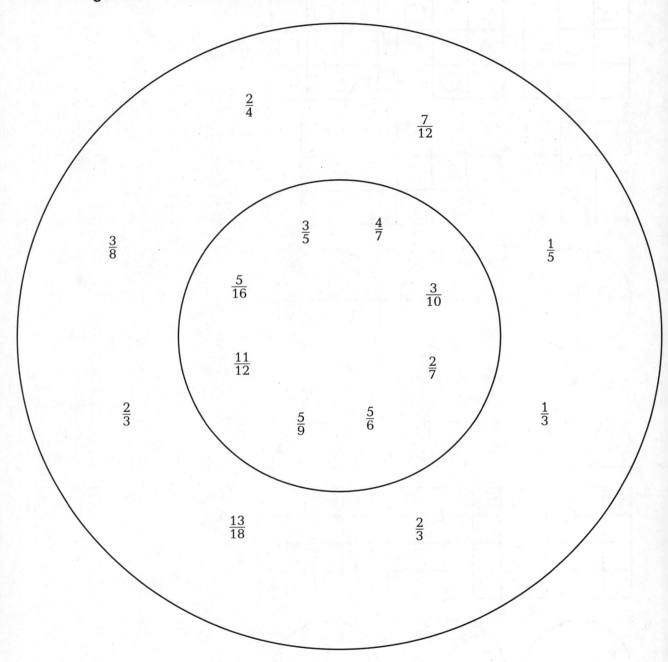

Describe the pattern. _____

Batter Up!

The bases are loaded, and the next batter comes up to the plate.

Each fraction represents a person on base, the batter, or the batter on deck. Starting with the batter on deck, next the batter, and then each base in order, the fractions should be from least to greatest. Help put them into the correct order.

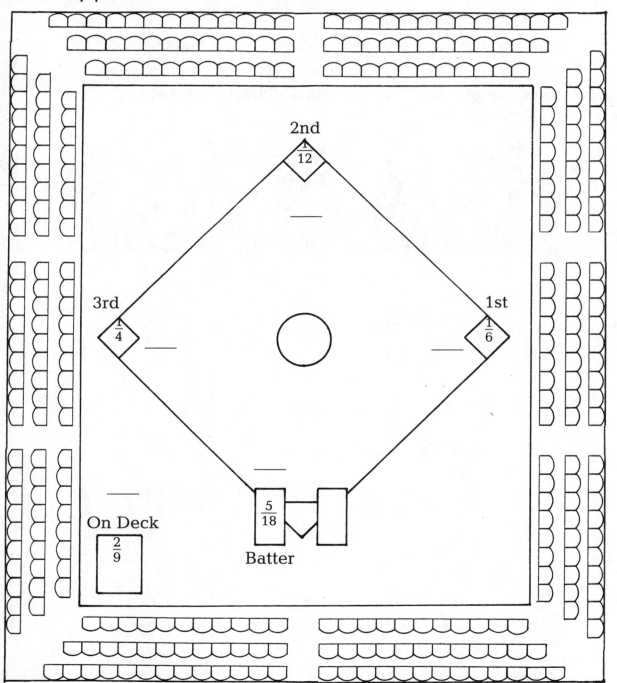

Fancy Footwork

A shoe salesman was at an outdoor party. On the way home he tried to remember what the other 24 people were wearing on their feet.

$\frac{1}{3}$ of the people had on white tennis shoes.

$\frac{1}{6}$ of the people had on sandals.

$\frac{1}{4}$ of the people had on boots.

Of the remaining people, $\frac{1}{2}$ had on high heels and $\frac{1}{2}$ had on no shoes.

Find how many people were in each group and draw what they were wearing on their feet.

Number Pyramids

The numbers in the pyramids are found by using one of these simple formulas:

$A + B = C$ or $C - A = B$ or $C - B = A$

If you know some of the numbers, you can find the rest.

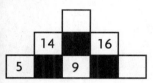

$14 + 16 = 30$

$16 - 9 = 7$

To find the top number, add.

To find the lower number, subtract.

Find the missing numbers in each pyramid.

1.

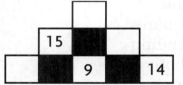

2.

3.

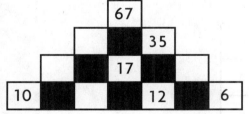

4.

Now, make your own number pyramids. Exchange them with a partner, and test each other's math skills.

Math Puzzles

Try these math puzzles. Have fun!

1. Try to figure out the pattern. What is the next line? Explain
the pattern.

     3 4 1 2 2

     4 5 2 0 0

     5 6 3 0 0

     6 7 4 2 8

     7 8 5 6 3 0

     _____

     _____

     _____

2. From the starting square, trace a path by moving one
square to a 1, then two squares to a 2, then three squares
to a 3, and so on. Your last move is eight squares, ending
on an 8. You may move horizontally, vertically, or both on
each move, but you may not move diagonally. There is
more than one solution.

Start	1	3	2	5	4	4	6
2	4	5	3	4	6	7	4
5	2	3	5	3	5	6	5
4	3	6	3	5	4	7	4
3	4	7	6	5	7	6	5
5	6	5	3	7	6	4	7
4	7	4	5	6	5	5	7
6	5	7	7	5	6	4	8

Equivalent Fractions Bingo

You can use your math skills to play bingo with equivalent fractions.

Materials   2 number cubes, markers

Rules   The object of the game is to be the first to cover a row of the gameboard with markers across, down, or diagonally.

1. Take turns rolling the 2 number cubes. Use one number as a numerator and one number as a denominator to make a fraction less than or equal to one.

2. Place a marker on the gameboard on each fraction you make or on a fraction that is equivalent to it.

For example, if you roll a 6 and a 2, you make the fraction $\frac{2}{6}$. This is not on the gameboard. An equivalent fraction for $\frac{2}{6}$ is $\frac{1}{3}$. So, you place a marker on the fraction $\frac{1}{3}$.

$\frac{1}{4}$	$\frac{3}{4}$	$\frac{1}{5}$	1	$\frac{2}{3}$
$\frac{2}{5}$	$\frac{1}{3}$	$\frac{1}{2}$	$\frac{1}{4}$	1
$\frac{1}{2}$	$\frac{1}{6}$	Free	$\frac{1}{3}$	$\frac{3}{4}$
$\frac{5}{6}$	1	$\frac{1}{4}$	$\frac{2}{3}$	$\frac{4}{5}$
$\frac{3}{4}$	$\frac{3}{5}$	$\frac{1}{3}$	$\frac{1}{2}$	$\frac{1}{6}$

Fix the Pattern

Each series of equivalent fractions creates a pattern. One fraction in each series does not follow the pattern. Cross out that fraction. Write the fraction that belongs in its place. The first one is done for you.

1. $\dfrac{2}{3}$ | $\cancel{\dfrac{4}{7}}$ | $\dfrac{6}{9}$ | $\dfrac{8}{12}$ | $\mathbf{\dfrac{4}{6}}$

2. $\dfrac{4}{16}$ | $\dfrac{2}{12}$ | $\dfrac{2}{8}$ | $\dfrac{1}{4}$ | _____

3. $\dfrac{2}{16}$ | $\dfrac{3}{24}$ | $\dfrac{4}{30}$ | $\dfrac{5}{40}$ | _____

4. $\dfrac{2}{4}$ | $\dfrac{4}{8}$ | $\dfrac{6}{16}$ | $\dfrac{16}{32}$ | _____

5. $\dfrac{3}{4}$ | $\dfrac{6}{8}$ | $\dfrac{10}{12}$ | $\dfrac{12}{16}$ | _____

6. $\dfrac{2}{7}$ | $\dfrac{4}{14}$ | $\dfrac{8}{21}$ | $\dfrac{16}{56}$ | _____

7. $\dfrac{25}{28}$ | $\dfrac{15}{21}$ | $\dfrac{10}{14}$ | $\dfrac{5}{7}$ | _____

8. $\dfrac{1}{4}$ | $\dfrac{3}{10}$ | $\dfrac{9}{36}$ | $\dfrac{27}{108}$ | _____

9. $\dfrac{20}{32}$ | $\dfrac{15}{24}$ | $\dfrac{10}{16}$ | $\dfrac{6}{8}$ | _____

10. $\dfrac{9}{11}$ | $\dfrac{18}{22}$ | $\dfrac{24}{33}$ | $\dfrac{36}{44}$ | _____

11. $\dfrac{4}{10}$ | $\dfrac{8}{20}$ | $\dfrac{10}{30}$ | $\dfrac{16}{40}$ | _____

12. $\dfrac{1}{6}$ | $\dfrac{3}{24}$ | $\dfrac{9}{72}$ | $\dfrac{27}{216}$ | _____

13. $\dfrac{1}{5}$ | $\dfrac{2}{15}$ | $\dfrac{3}{15}$ | $\dfrac{4}{20}$ | _____

14. $\dfrac{7}{9}$ | $\dfrac{15}{18}$ | $\dfrac{21}{27}$ | $\dfrac{28}{36}$ | _____

15. $\dfrac{27}{81}$ | $\dfrac{12}{27}$ | $\dfrac{3}{9}$ | $\dfrac{1}{3}$ | _____

16. $\dfrac{1}{2}$ | $\dfrac{2}{4}$ | $\dfrac{3}{5}$ | $\dfrac{4}{8}$ | _____

Calculating Fractions

You can use a calculator to see if two fractions are equivalent.

Use your calculator to divide the numerator by the denominator in the two fractions. Round the answer to the nearest hundredth where necessary. Does $\frac{1}{4} = \frac{2}{8}$?

1 $\boxed{\div}$ 4 $\boxed{=}$ 0.25 $\longleftarrow$ $\quad$ $\frac{1}{4} = 0.25$

$\qquad$ equal

2 $\boxed{\div}$ 8 $\boxed{=}$ 0.25 $\longleftarrow$ $\quad$ $\frac{2}{8} = 0.25$

So, $\frac{1}{4} = \frac{2}{8}$. They are equivalent.

Are the fractions equivalent? Use a calculator to find out.
Write *yes* or *no*.

1. $\frac{6}{8} =$ _____

$\frac{30}{40} =$ _____ _____

2. $\frac{7}{9} =$ _____

$\frac{77}{99} =$ _____ _____

3. $\frac{5}{6} =$ _____

$\frac{15}{16} =$ _____ _____

4. $\frac{1}{2} =$ _____

$\frac{49}{98} =$ _____ _____

5. $\frac{17}{20} =$ _____

$\frac{45}{50} =$ _____ _____

6. $\frac{14}{21} =$ _____

$\frac{22}{33} =$ _____ _____

7. $\frac{8}{17} =$ _____

$\frac{48}{102} =$ _____ _____

8. $\frac{58}{64} =$ _____

$\frac{13}{16} =$ _____ _____

9. $\frac{3}{4} =$ _____

$\frac{66}{88} =$ _____ _____

10. $\frac{7}{8} =$ _____

$\frac{126}{152} =$ _____ _____

11. $\frac{3}{5} =$ _____

$\frac{66}{110} =$ _____ _____

12. $\frac{16}{20} =$ _____

$\frac{66}{80} =$ _____ _____

13. $\frac{2}{3} =$ _____

$\frac{28}{45} =$ _____ _____

14. $\frac{10}{21} =$ _____

$\frac{190}{399} =$ _____ _____

Riddle Time

What has locks but no keys? To find out, write each fraction in simplest form. Then match the letter with the fraction in one of the boxes at the bottom of the page. These boxes will spell out the answer.

1. $\dfrac{15}{25}$ = _____ A

2. $\dfrac{4}{5}$ = _____ H

3. $\dfrac{8}{64}$ = _____ C

4. $\dfrac{45}{120}$ = _____ M

5. $\dfrac{36}{48}$ = _____ A

6. $\dfrac{25}{40}$ = _____ A

7. $\dfrac{20}{50}$ = _____ L

8. $\dfrac{4}{12}$ = _____ T

9. $\dfrac{100}{120}$ = _____ E

10. $\dfrac{16}{64}$ = _____ A

11. $\dfrac{5}{25}$ = _____ N

12. $\dfrac{12}{24}$ = _____ A

13. $\dfrac{16}{24}$ = _____ N

14. $\dfrac{77}{88}$ = _____ P

$\dfrac{1}{3}$	$\dfrac{4}{5}$	$\dfrac{5}{6}$	$\dfrac{7}{8}$	$\dfrac{3}{5}$	$\dfrac{2}{3}$	$\dfrac{1}{4}$	$\dfrac{3}{8}$	$\dfrac{1}{2}$

$\dfrac{1}{8}$	$\dfrac{3}{4}$	$\dfrac{1}{5}$	$\dfrac{5}{8}$	$\dfrac{2}{5}$

Way to Go!

Follow each path and find the sum. Put a star at the end of the
path that has the largest sum.

Path 1

$\frac{1}{4} \rightarrow \frac{2}{4} \rightarrow \frac{3}{4} \rightarrow \frac{1}{4} \rightarrow \frac{2}{4} \rightarrow \frac{1}{4} \rightarrow \frac{3}{4} \rightarrow \frac{1}{4} \rightarrow \frac{2}{4} \rightarrow \frac{3}{4} \rightarrow \frac{1}{4} \rightarrow \frac{2}{4} \rightarrow \frac{1}{4} \rightarrow$

$\frac{3}{4} \rightarrow \frac{1}{4} \rightarrow \frac{1}{4} \rightarrow \frac{1}{4} \rightarrow \frac{3}{4} \rightarrow \frac{2}{4} \rightarrow \frac{1}{4} \rightarrow \frac{1}{4}$

_____

Path 2

$\frac{1}{3} \rightarrow \frac{2}{3} \rightarrow \frac{1}{3} \rightarrow \frac{2}{3} \rightarrow \frac{2}{3} \rightarrow \frac{2}{3} \rightarrow \frac{1}{3} \rightarrow \frac{2}{3} \rightarrow \frac{2}{3} \rightarrow \frac{2}{3} \rightarrow \frac{2}{3} \rightarrow \frac{2}{3} \rightarrow \frac{2}{3} \rightarrow$

$\frac{2}{3} \rightarrow \frac{1}{3} \rightarrow \frac{1}{3}$

_____

Path 3

$\frac{2}{5} \rightarrow \frac{3}{5} \rightarrow \frac{4}{5} \rightarrow \frac{2}{5} \rightarrow \frac{4}{5} \rightarrow \frac{1}{5} \rightarrow \frac{4}{5} \rightarrow \frac{3}{5} \rightarrow \frac{4}{5} \rightarrow \frac{1}{5} \rightarrow \frac{2}{5} \rightarrow \frac{1}{5} \rightarrow \frac{4}{5} \rightarrow$

$\frac{1}{5} \rightarrow \frac{4}{5} \rightarrow \frac{3}{5} \rightarrow \frac{2}{5} \rightarrow \frac{1}{5} \rightarrow \frac{1}{5} \rightarrow \frac{3}{5}$

_____

Path 4

$\frac{1}{6} \rightarrow \frac{1}{6} \rightarrow \frac{1}{6} \rightarrow \frac{2}{6} \rightarrow \frac{3}{6} \rightarrow \frac{3}{6} \rightarrow \frac{2}{6} \rightarrow \frac{5}{6} \rightarrow \frac{1}{6} \rightarrow \frac{1}{6} \rightarrow \frac{1}{6} \rightarrow \frac{2}{6} \rightarrow \frac{2}{6} \rightarrow$

$\frac{1}{6} \rightarrow \frac{3}{6} \rightarrow \frac{1}{6} \rightarrow \frac{1}{6} \rightarrow \frac{1}{6} \rightarrow \frac{2}{6} \rightarrow \frac{1}{6} \rightarrow \frac{1}{6} \rightarrow \frac{1}{6} \rightarrow \frac{3}{6} \rightarrow \frac{2}{6}$

_____

Mystery Fraction

Pick a fraction from the list below. Write that fraction in the circle that has exact space for it. Continue until you fill both circles. Then answer the questions below.

One fraction, the mystery fraction, fits into both circles. All other fractions fit into only one circle.

$$\frac{1}{8} \qquad \frac{1}{4} \qquad \frac{1}{12} \qquad \frac{3}{8} \qquad \frac{3}{16} \qquad \frac{6}{12} \qquad \frac{1}{16} \qquad \frac{1}{6}$$

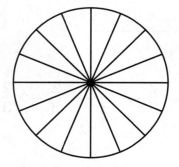

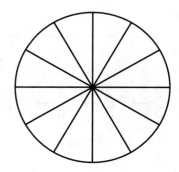

Circle 1
(sixteenths)

Circle 2
(twelfths)

1. What is the mystery fraction? _____

2. Write the fractions that fill circle 1. _____

   What is their sum? _____

3. Write the fractions that fill circle 2. _____

   What is their sum? _____

4. What is the smallest fraction listed? _____

5. What is the largest fraction listed? _____

Fraction Fill-in

Amanda is given a square like the one here. She is asked to fill in each box with one of the following fractions:

$$\frac{1}{6}, \frac{4}{9}, \frac{7}{12}, \frac{1}{18}, \frac{5}{18}, \frac{31}{36}$$

She is told to follow these rules:

- The sum of Row 1 is $\frac{31}{36}$.
- The sum of Row 2 is $\frac{11}{18}$.
- The sum of Column 1 is $1\frac{1}{36}$.
- The sum of Column 2 is $\frac{4}{9}$.

	Column 1	Column 2
Row 1	_____	_____
Row 2	_____	_____

1. What does her square look like?

Joel is given a square like the one here. He is asked to fill in each box with one of the following fractions:

$$\frac{5}{24}, \frac{1}{7}, \frac{3}{8}, \frac{5}{9}, \frac{4}{12}, \frac{1}{16}$$

He is told to follow these rules:

- The sum of Row 1 is $\frac{7}{12}$.
- The sum of Row 2 is $\frac{19}{48}$.
- The sum of Column 1 is $\frac{7}{16}$.
- The sum of Column 2 is $\frac{13}{24}$.

	Column 1	Column 2
Row 1	_____	_____
Row 2	_____	_____

2. What does his square look like?

The Fraction Bingo

The students in Mrs. Stern's class love to play Fraction Bingo. Here's how they play it.

- Each student gets a card with different fractions on it.

- Mrs. Stern calls out the fractions, one at a time.

- The first student to get three in a row—horizontally, vertically, or diagonally—wins.

In today's game, the sum of the winning combination is $3\frac{2}{15}$. Find the sums of the rows of three fractions.

Who won Fraction Bingo? _____

What was the winning combination? _____

$\frac{1}{2}$	$\frac{3}{5}$	$\frac{8}{15}$
$\frac{9}{14}$	free	$\frac{3}{6}$
$\frac{6}{7}$	$\frac{11}{35}$	$\frac{4}{5}$

Phil

$\frac{2}{15}$	$\frac{4}{3}$	$\frac{7}{2}$
$\frac{8}{3}$	free	$\frac{6}{8}$
$\frac{3}{5}$	$\frac{5}{8}$	$\frac{12}{16}$

Juan

$\frac{1}{5}$	$\frac{6}{35}$	$\frac{3}{7}$
$\frac{2}{3}$	free	$\frac{3}{4}$
$\frac{5}{6}$	$\frac{4}{5}$	$\frac{3}{2}$

Yvette

$\frac{2}{9}$	$\frac{2}{3}$	$\frac{3}{2}$
$\frac{1}{2}$	free	$\frac{3}{6}$
$\frac{5}{6}$	$\frac{1}{4}$	$\frac{1}{12}$

Miki

Model Fractions

Place one fraction from the list below in each circle. Arrange the fractions so that the sum of each diagonal is the same.

$$\frac{1}{2}, \frac{1}{3}, \frac{2}{3}, \frac{3}{4}, \frac{3}{6}, \frac{5}{8}, \frac{1}{12}, \frac{4}{12}, \frac{4}{24}$$

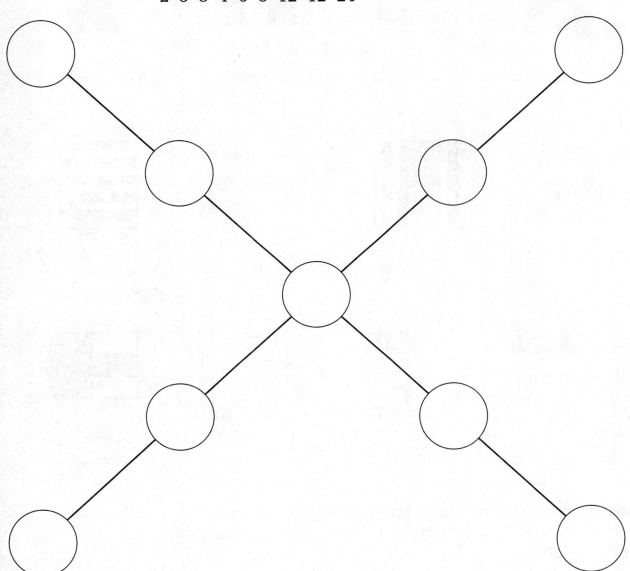

What is the sum of each diagonal? _____

Like What?

The shaded part of each figure represents a fraction.
Write the fraction.

1.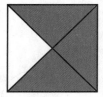

_____

2.

_____

3.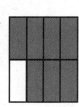

_____

4.

_____

5.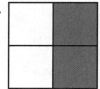

_____

6.

_____

7.

_____

8.

_____

9.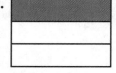

_____

10.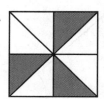

_____

11.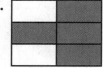

_____

12.

_____

Find the figures that represent fractions with the same
denominators. Write a subtraction sentence for each pair of like
fractions. Solve.

13. _____

14. _____

15. _____

16. _____

17. _____

18. _____

Cut Me Up!

You can subtract only like fractions.

Find $\frac{1}{2} - \frac{1}{4}$.

 —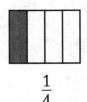

$$\frac{1}{2} \qquad\qquad \frac{1}{4}$$

Divide each half of the first figure in half. Both figures now have equal parts. Subtract the like fractions.

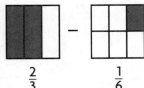

$$\frac{2}{4} \qquad\qquad \frac{1}{4} \qquad\qquad \frac{1}{4}$$

So, $\frac{1}{2} - \frac{1}{4} = \frac{1}{4}$.

For each pair of figures, find a way to divide them so that they have equal parts. Explain. Then subtract.

1.

$$\frac{2}{3} \qquad\qquad \frac{1}{6}$$

_____

_____

2.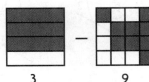

$$\frac{3}{4} \qquad\qquad \frac{9}{16}$$

_____

_____

3.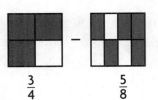

$$\frac{3}{4} \qquad\qquad \frac{5}{8}$$

_____

_____

4.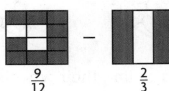

$$\frac{9}{12} \qquad\qquad \frac{2}{3}$$

_____

_____

Odd Wheels Keep On Turning

Paul, Lin, and Albert are rolling their favorite shape wheels up a hill. All of the wheels begin with side 1 on the ground.

They all roll their wheels one time.

They all roll their wheels a second time.

Then they all roll their wheels a third time.

They all keep on rolling their wheels, one side at a time.

Albert figures out that they will reach the top of the hill the next time that all of the shapes have side 1 on the ground again. How many times will they need to roll their shapes? HINT: One strategy you may use is to find the LCD.

_____

New Rulers for Sale

Have you ever wanted to measure $\frac{1}{3}$ inch?
Well, here is the ruler for you. It is the Thirds Ruler.

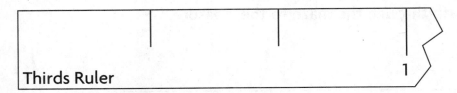

Thirds Ruler   1

1. Label the marks on the ruler. List them. _____

2. Make a mark halfway between each pair of marks on the ruler.

3. Label the new marks you made on the ruler. List them. _____

4. Make a mark halfway between each mark you now have on the ruler.

5. Label the new marks you made on the ruler. List them. _____

Use your Thirds Ruler to find the difference.

6. $\frac{3}{6} - \frac{1}{3} =$ _____

7. $\frac{9}{12} - \frac{4}{6} =$ _____

8. $\frac{11}{12} - \frac{2}{3} =$ _____

Here is a Fifths Ruler.

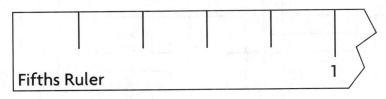

Fifths Ruler   1

9. Label the marks on the ruler. List them. _____

10. Make a mark halfway between each pair of marks on the ruler.

11. Label the new marks you made on the ruler. List them. _____

Use your Fifths Ruler to find the difference.

12. $\frac{9}{10} - \frac{3}{5} =$ _____

13. $\frac{4}{5} - \frac{3}{10} =$ _____

14. $\frac{7}{10} - \frac{1}{2} =$ _____

Forward and Backward

Sometimes you have to go backward to go forward. Do both to solve the maze below.

Begin at START. Find a path through the maze to the treasure.

START

Guess Where I Go

Is each fraction closest to 0, to $\frac{1}{2}$, or to 1?

Write 0, $\frac{1}{2}$, or 1 on the line.

Then write the fraction in the correct circle.

1. $\frac{12}{25}$ 2. $\frac{21}{40}$ 3. $\frac{8}{17}$ 4. $\frac{13}{30}$

_____   _____   _____   _____

5. $\frac{18}{19}$ 6. $\frac{22}{24}$ 7. $\frac{3}{16}$ 8. $\frac{9}{11}$

_____   _____   _____   _____

9. $\frac{1}{12}$ 10. $\frac{2}{31}$ 11. $\frac{15}{26}$ 12. $\frac{17}{22}$

_____   _____   _____   _____

0   **$\frac{1}{2}$**   **1**

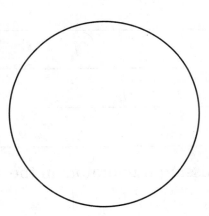 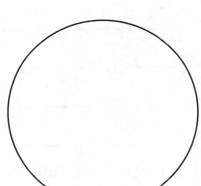

13. How many fractions are closest to 0?

14. How many fractions are closest to $\frac{1}{2}$?

15. How many fractions are closest to 1?

_____   _____   _____

16. Make up your own fractions, and tell which circle they belong in.

_____

Missing Parts

Some of the numerators are missing from the fractions below.
Can you find them?

1. $\dfrac{4}{10} + \dfrac{\Box}{10} = \dfrac{3}{5}$

2. $\dfrac{\Box}{35} - \dfrac{7}{35} = \dfrac{1}{7}$

3. $\dfrac{11}{18} + \dfrac{\Box}{18} = \dfrac{8}{9}$

4. $\dfrac{12}{81} - \dfrac{\Box}{81} = \dfrac{1}{9}$

5. $\dfrac{\Box}{48} + \dfrac{5}{48} = \dfrac{1}{2}$

6. $\dfrac{\Box}{20} + \dfrac{7}{20} = \dfrac{3}{4}$

7. $\dfrac{\Box}{36} - \dfrac{8}{36} = \dfrac{3}{4}$

8. $\dfrac{\Box}{32} + \dfrac{5}{32} = \dfrac{3}{8}$

9. $\dfrac{14}{27} - \dfrac{\Box}{27} = \dfrac{1}{3}$

10. $\dfrac{8}{21} + \dfrac{\Box}{21} = \dfrac{2}{3}$

11. $\dfrac{\Box}{49} - \dfrac{21}{49} = \dfrac{2}{7}$

12. $\dfrac{66}{77} - \dfrac{\Box}{77} = \dfrac{1}{7}$

13. How did you find the missing numerators in the addition problems?

_____

_____

_____

_____

14. How did you find the missing numerators in the subtraction problems?

_____

_____

_____

_____

Hal's Hat Store

Find the sum or difference of the fractions in each hat. Write your answers in simplest form. Then color the hats to match the table of denominators below.

$$\frac{6}{9}$$
$$-\frac{1}{3}$$
_____

$$\frac{2}{5}$$
$$+\frac{1}{10}$$
_____

$$\frac{5}{12}$$
$$-\frac{1}{6}$$
_____

$$\frac{11}{12}$$
$$-\frac{4}{12}$$
_____

$$\frac{3}{4}$$
$$-\frac{1}{12}$$
_____

$$\frac{2}{9}$$
$$+\frac{2}{3}$$
_____

$$\frac{3}{6}$$
$$+\frac{1}{4}$$
_____

$$\frac{6}{10}$$
$$-\frac{2}{10}$$
_____

$$\frac{1}{5}$$
$$+\frac{2}{5}$$
_____

1. How many hats are purple?

_____

2. How many hats have answers that could have equivalent fractions with a denominator of 12?

_____

3. List the answers that are in these hats.

_____

Denominator of the Answer	Color
2	pink
3	green
4	red
5	blue
9	brown
12	purple

The Race Is On!

Mount Sierra Elementary School is having its annual race day.
Solve the problem in the first hurdle. Write the answer in
simplest form. Then write that answer in the second hurdle,
and solve the problem. Continue in this way to the finish line.
Get ready! Get set! Go!

1. START

$$\frac{3}{4} - \frac{1}{8} = \underline{\quad}$$

$$\underline{\quad} - \frac{1}{2} = \underline{\quad}$$

$$\underline{\quad} + \underline{\quad} = \frac{5}{16}$$

 FINISH

2. START

$$\frac{5}{12} + \frac{1}{4} = \underline{\quad}$$

$$\underline{\quad} - \frac{1}{2} = \underline{\quad}$$

$$\underline{\quad} + \underline{\quad} = \frac{5}{18}$$

 FINISH

3. START

$$\frac{1}{2} - \frac{3}{10} = \underline{\quad}$$

$$\underline{\quad} + \frac{1}{3} = \underline{\quad}$$

$$\underline{\quad} + \underline{\quad} = \frac{2}{3}$$

 FINISH

4. START

$$\frac{5}{6} - \frac{2}{3} = \underline{\quad}$$

$$\underline{\quad} + \frac{1}{4} = \underline{\quad}$$

$$\underline{\quad} + \underline{\quad} = \frac{3}{4}$$

 FINISH

Busy Studying

Ms. Jacks gave a lot of homework this weekend. The table shows how much of the day some of her students spent on it. Tell whether you would add or subtract to solve each problem. Use the table to solve.

Name	Time Spent on Homework
Mark	$\frac{3}{4}$ of the day
Lisa	$\frac{3}{8}$ of the day
Pete	$\frac{1}{8}$ of the day
Chris	$\frac{5}{8}$ of the day
Theresa	$\frac{1}{2}$ the day
Charlie	$\frac{1}{4}$ of the day

1. How much more of the day did Mark spend on homework than Theresa?

_____

2. Who spent the greatest fraction of the day working? the least? What is the difference between the two fractions?

_____

_____

3. How many days' worth of time did Lisa, Pete, and Chris spend in all on homework?

_____

4. Charlie and Lisa started their homework at the same time on Sunday. Who finished first?

_____

5. Use the table to make up your own addition problem.

_____

_____

_____

_____

6. Use the table to make up your own subtraction problem.

_____

_____

_____

_____

Got It?

Find each sum or difference below. Write the answer in simplest form. In the matching boxes below, write the letters that are next to the answers. The message will answer the question.

1. $\dfrac{14}{27} + \dfrac{4}{9} =$ _____ O

2. $\dfrac{3}{4} + \dfrac{3}{16} =$ _____ E

3. $\dfrac{4}{5} - \dfrac{1}{15} =$ _____ R

4. $\dfrac{7}{12} + \dfrac{7}{36} =$ _____ T

5. $\dfrac{5}{6} - \dfrac{3}{5} =$ _____ F

6. $\dfrac{1}{8} + \dfrac{5}{6} =$ _____ E

7. $\dfrac{4}{7} + \dfrac{1}{3} =$ _____ H

8. $\dfrac{7}{15} - \dfrac{3}{10} =$ _____ H

9. $\dfrac{2}{5} - \dfrac{1}{8} =$ _____ T

10. $\dfrac{2}{5} - \dfrac{1}{4} =$ _____ H

11. $\dfrac{5}{7} - \dfrac{3}{5} =$ _____ L

12. $\dfrac{1}{2} + \dfrac{2}{9} =$ _____ A

What most resembles half a pear?

$\dfrac{7}{9}$	$\dfrac{3}{20}$	$\dfrac{23}{24}$	$\dfrac{26}{27}$	$\dfrac{11}{40}$	$\dfrac{19}{21}$	$\dfrac{15}{16}$	$\dfrac{11}{15}$	$\dfrac{1}{6}$	$\dfrac{13}{18}$	$\dfrac{4}{35}$	$\dfrac{7}{30}$

Rules Rule

Use your ruler. Measure each line segment to the nearest $\frac{1}{4}$ inch.
Exercise 1 is done for you.

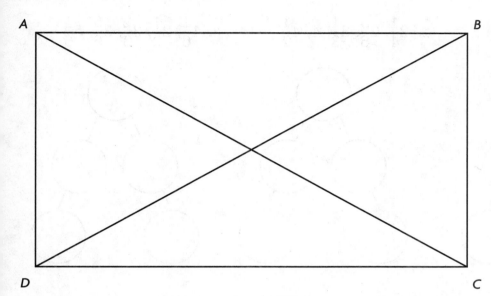

1. $AB =$ $4\frac{1}{2}$ in. _____

2. $BC =$ _____

3. $DC =$ _____

4. $AD =$ _____

5. $AC =$ _____

6. $BD =$ _____

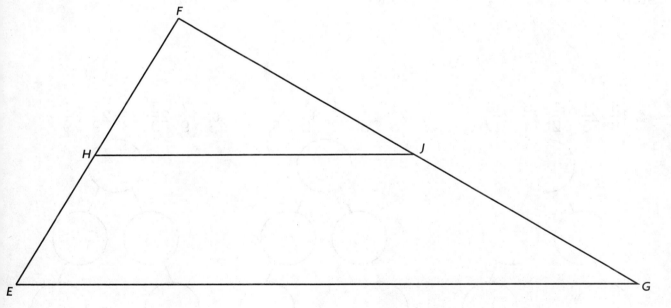

7. $FE =$ _____   8. $HE =$ _____

9. $FG =$ _____   10. $FJ =$ _____

11. $EG =$ _____   12. $JG =$ _____

13. $FH =$ _____   14. $HJ =$ _____

And the Answer Is

For each problem below, arrange the numbers in the circles so
that both diagonals have the same sum.

1. $\frac{1}{2}$, $2\frac{3}{8}$, $3\frac{1}{4}$, $4\frac{1}{2}$, $5\frac{1}{8}$

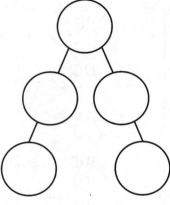

Sum = ————

2. $2\frac{1}{4}$, $2\frac{5}{6}$, $3\frac{1}{3}$, 4, $4\frac{1}{2}$

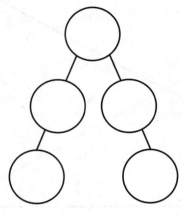

Sum = ————

3. $1\frac{1}{6}$, $2\frac{1}{2}$, $3\frac{2}{3}$, $3\frac{5}{6}$, $5\frac{1}{6}$

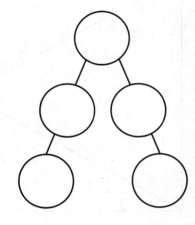

Sum = ____

4. $1\frac{5}{8}$, $2\frac{1}{3}$, $2\frac{1}{2}$, $3\frac{3}{4}$, $3\frac{11}{12}$

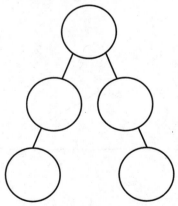

Sum = ____

5. $2\frac{1}{5}$, $2\frac{3}{5}$, $2\frac{1}{2}$, $3\frac{1}{2}$, $3\frac{3}{5}$

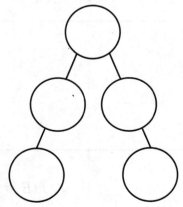

Sum = ____

6. $1\frac{7}{9}$, $2\frac{2}{9}$, $2\frac{1}{3}$, $2\frac{7}{9}$, $3\frac{1}{9}$

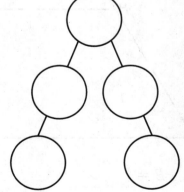

Sum = ____

Subtraction Madness

Fill each △ with the correct mixed number: $2\frac{1}{6}$, $5\frac{5}{6}$, $2\frac{1}{2}$, or $5\frac{1}{3}$.

1.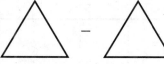

2.

Fill each □ with the correct mixed number: $4\frac{7}{8}$, $4\frac{3}{8}$, $1\frac{1}{2}$, or $1\frac{1}{4}$.

3.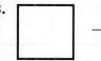

4.

Fill each ◇ with the correct mixed number: $3\frac{2}{5}$, $6\frac{3}{5}$, $6\frac{7}{10}$, or $3\frac{1}{10}$.

5.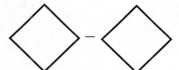

6.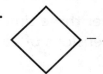

Fill each ⬠ with the correct mixed number: $2\frac{1}{4}$, $1\frac{1}{12}$, $4\frac{1}{2}$, or $5\frac{1}{2}$.

7.

8.

Fill each ⬡ with the correct mixed number: $6\frac{7}{9}$, $2\frac{1}{3}$, $3\frac{1}{3}$, or $5\frac{3}{4}$.

9.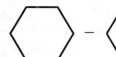

10.

Unfolding!

The piece of paper below was folded three times. Unfolding it shows its original dimensions.

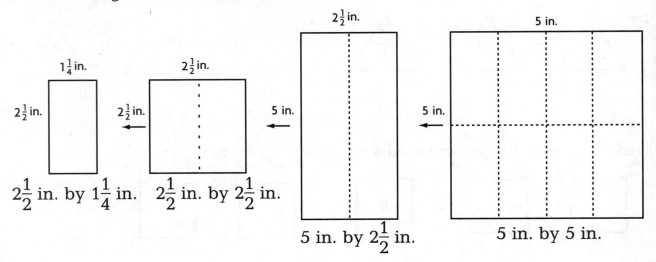

$2\frac{1}{2}$ in. by $1\frac{1}{4}$ in. $2\frac{1}{2}$ in. by $2\frac{1}{2}$ in.   5 in. by $2\frac{1}{2}$ in.   5 in. by 5 in.

The original dimensions are 5 in. by 5 in.

Unfold each piece of paper three times to find the original dimensions. Give the dimensions of each unfolding.

1.

$3\frac{1}{2}$ in. $2\frac{1}{2}$ in. ⟵   ⟵   ⟵

_____   _____   _____

The original dimensions are _____

2.

$1\frac{1}{8}$ in. $1\frac{1}{8}$ in. ⟵   ⟵   ⟵

_____   _____   _____

The original dimensions are ⟶ _____

Add or Subtract?

Decide if you should add or subtract each mixed number to go
from the start to the finish. Put + or − in each circle.

1. Start   Finish

$5\frac{3}{4}$ ◯ $1\frac{1}{2}$ ◯ $2\frac{1}{4}$ ◯ $3\frac{3}{8}$ = $8\frac{3}{8}$

2. Start   Finish

$7\frac{2}{3}$ ◯ $3\frac{1}{6}$ ◯ $1\frac{1}{4}$ ◯ $1\frac{3}{8}$ = $4\frac{5}{8}$

3. Start   Finish

$2\frac{3}{9}$ ◯ $1\frac{1}{9}$ ◯ $4\frac{4}{9}$ ◯ $3\frac{1}{6}$ = $2\frac{1}{2}$

4. Start   Finish

$10\frac{11}{12}$ ◯ $3\frac{3}{4}$ ◯ $4\frac{1}{3}$ ◯ $1\frac{1}{6}$ = $1\frac{2}{3}$

5. Start   Finish

$7\frac{4}{5}$ ◯ $2\frac{1}{10}$ ◯ $3\frac{2}{5}$ ◯ $4\frac{3}{5}$ = $4\frac{1}{2}$

6. Start   Finish

$3\frac{1}{3}$ ◯ $2\frac{1}{4}$ ◯ $1\frac{1}{6}$ ◯ $2\frac{1}{2}$ = $4\frac{1}{4}$

7. Start   Finish

$11\frac{7}{9}$ ◯ $2\frac{2}{3}$ ◯ $1\frac{1}{3}$ ◯ $7\frac{1}{9}$ = $3\frac{1}{3}$

Uncustomary Units of Measure

Throughout history, people have used different units of measure. You can even use some of today's sports greats as units of measure.

Gheorghe Muresan, the tallest player in the NBA, is 7 feet 7 inches tall.

Basketball star Michael Jordan is 6 feet 6 inches tall.

Figure skater Michelle Kwan is 5 feet 2 inches tall.

Use the sports figures above as units of measure. (Round all figures and answers to the nearest half foot.)

1. The Washington Monument is 555 ft 5 in. tall.

   a. About how many Gheorghe Muresan units tall is the Washington Monument? _____

   b. about how many Michael Jordan units? _____

   c. about how many Michelle Kwan units? _____

2. The Empire State Building is 1,472 ft tall.

   a. About how many Gheorghe Muresan units tall is the Empire State Building? _____

   b. about how many Michael Jordan units? _____

   c. about how many Michelle Kwan units? _____

3. The Great Pyramid of Khufu in Giza, Egypt, is 482 ft tall.

   a. About how many Gheorghe Muresan units tall is the Great Pyramid? _____

   b. about how many Michael Jordan units? _____

   c. about how many Michelle Kwan units? _____

A Customary Message

After each unit of measure, write the letter of its equivalent measure. You may use a calculator.

1. 5 yd _____   A. 3 yd

2. 10 ft _____   O. 9 yd

3. 8 ft 12 in. _____   C. 300 ft

4. 4 yd 6 ft _____   H. 360 in.

5. 7 yd _____   E. 3,520 yd

6. 100 yd _____   Q. 220 in.

7. 1,320 ft _____   K. 33 yd 1 ft

8. 10 yd _____   D. 180 in.

9. 2 mi _____   P. 3 mi

10. 100 ft _____   J. 21 ft

11. 120 in. _____   G. 3 yd 1 ft

12. 324 in. _____   N. 26.2 mi

13. 6 yd 4 in. _____   B. 440 yd

14. 2,640 ft _____   I. $\frac{1}{2}$ mi

15. 46,112 yd _____   F. 18 ft

16. 5,280 yd _____   R. 150 ft

17. 50 yd _____   T. 10 ft

Find the message by filling in the correct letter above each number.

___ ___ ___ ___ ___   ___ ___ ___ !
 2   17   9   3   11   5   12   7

It Doesn't Add Up

Each of the following exercises has an error. Your mission,
should you wish to accept it, is to find and correct the error.
The answer is always correct, so the error is above the line.
The first one is done for you.

1.
$$
\begin{array}{ll}
8 \text{ ft} & 4 \text{ in.} \quad \cancel{5 \text{ in.}} \\
+ 2 \text{ ft} & 10 \text{ in.} \\
\hline
11 \text{ ft} & 2 \text{ in.}
\end{array}
$$

2.
$$
\begin{array}{ll}
7 \text{ yd} & 4 \text{ ft} \\
- 4 \text{ yd} & 5 \text{ ft} \\
\hline
3 \text{ yd} & 2 \text{ ft}
\end{array}
$$

3.
$$
\begin{array}{ll}
10 \text{ yd} & 1 \text{ ft} \\
- 7 \text{ yd} & 2 \text{ ft} \\
\hline
1 \text{ yd} & 2 \text{ ft}
\end{array}
$$

4.
$$
\begin{array}{ll}
26 \text{ yd} & 5 \text{ ft} \\
+ 19 \text{ yd} & 6 \text{ ft} \\
\hline
49 \text{ yd} & 1 \text{ ft}
\end{array}
$$

5.
$$
\begin{array}{ll}
10 \text{ ft} & 7 \text{ in.} \\
- 9 \text{ ft} & 8 \text{ in.} \\
\hline
2 \text{ ft} & 11 \text{ in.}
\end{array}
$$

6.
$$
\begin{array}{ll}
36 \text{ ft} & 11 \text{ in.} \\
+ 10 \text{ ft} & 6 \text{ in.} \\
\hline
46 \text{ ft} & 5 \text{ in.}
\end{array}
$$

7.
$$
\begin{array}{ll}
8 \text{ ft} & 4 \text{ in.} \\
- 5 \text{ ft} & 6 \text{ in.} \\
\hline
4 \text{ ft} & 10 \text{ in.}
\end{array}
$$

8.
$$
\begin{array}{ll}
7 \text{ yd} & 2 \text{ ft} \\
+ 7 \text{ yd} & 2 \text{ ft} \\
\hline
14 \text{ yd} & 1 \text{ ft}
\end{array}
$$

9.
$$
\begin{array}{ll}
19 \text{ yd} & 2 \text{ ft} \\
- 13 \text{ yd} & 2 \text{ ft} \\
\hline
5 \text{ yd} & 2 \text{ ft}
\end{array}
$$

10.
$$
\begin{array}{ll}
12 \text{ ft} & 3 \text{ in.} \\
+ 8 \text{ ft} & 11 \text{ in.} \\
\hline
20 \text{ ft} & 2 \text{ in.}
\end{array}
$$

Common Bond

Describe what the numbers in each exercise have in common.

1.   34   45   78   123   678   89

_____

_____

2.   34   43   223   70   61   52   16
         1222   331   1111111

_____

_____

3.   46   68   97   20   53   75   31

_____

_____

4.   343   6,556   12,121   66   95,059   38,883

_____

_____

5.   3,217   5,153   2,189   362   482   177   5,408

_____

_____

6.   927   413   734   211   321   835

_____

_____

7.   7321   4520   6848   7963   3412   5525
                   2816   5315   9981

_____

_____

Atlas Stones

At the annual "World's Strongest Person" competition, no event tests athletic strength better than the Stones of Atlas. Competitors must lift six progressively larger round stones onto 3-foot platforms. The stones are huge—about 2–3 feet in diameter. Their weight is staggering.

The weight of the Stones of Atlas is given in the ancient measurement of *stones*. A stone is about 13.5 pounds.

Convert the weight of the 6 Atlas Stones into pounds.

1. 10 stones = _____ lb

2. 13 stones = _____ lb

3. 15 stones = _____ lb

4. 18 stones = _____ lb

5. 20 stones = _____ lb

6. 23 stones = _____ lb

7. In the 1995 event, one competitor executed a dead lift of 960 pounds. How many stones would that be?

_____

8. Some of the competitors in the "World's Strongest Person" competition weigh 30 stones. What is their weight in pounds?

_____

9. Figure out how much the following people in Doreen's family weigh in stones. Complete the chart. Round to the nearest tenth.

Name	Weight in Pounds	Weight in Stones
Doreen	76	_____
Natalie	92	_____
Jake	105	_____
Mrs. Snell	146	_____
Mr. Snell	207	_____

In the Zone

The world is divided into 24 time zones, reflecting the Earth's rotation. When it's 8 A.M. in your zone, it's 8:00 P.M. on the other side of the Earth. When it is noon in New York, the time in other parts of the world is as follows:

| Hawai'i 8:00 A.M. | Seattle 9:00 A.M. | Denver 10:00 A.M. | Houston 11:00 A.M. |

| Spain 6:00 P.M. | Israel 7:00 P.M. | China 11:00 P.M. | New Zealand 5:00 A.M. |

Use the clocks above to answer the following questions.

1. When it is noon in Spain, what time is it in

   a. Denver? _____

   b. New Zealand? _____

   c. China? _____

2. When it is noon in Hawai'i, what time is it in

   a. Houston? _____

   b. Spain? _____

   c. Israel? _____

3. When it is 6:25 A.M. in Israel, what time is it in

   a. Seattle? _____

   b. China? _____

   c. Hawai'i? _____

One Life to Live

The life span of the average American is now about 76 years.

Complete the table. Use a calculator.

Average American Life Span	
Number of years	76
Number of months	_____
Number of days	_____
Number of hours	_____
Number of minutes	_____
Number of seconds	_____

Calculate how long you have lived to seconds. Round your age in years to the nearest month. Round your answers to the nearest whole number.

My Life So Far	
Number of years	_____
Number of months	_____
Number of days	_____
Number of hours	_____
Number of minutes	_____
Number of seconds	_____

Converting Temperatures

Air temperature is measured by a mercury thermometer. As the temperature rises, the mercury expands and rises in the thermometer tube. As the temperature falls, the mercury contracts and falls.

On the Fahrenheit scale, water freezes at 32°F and boils at 212°F. On the Celsius scale, water freezes at 0°C and boils at 100°C.

You can convert degrees Celsius (°C) to degrees Fahrenheit (°F) using a special formula:

For example, use the formula to convert 37°C to °F:

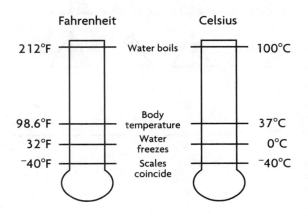

$(1.8 \times °C) + 32 = °F$
$(1.8 \times 37) + 32 =$
$\quad\quad 66.6 + 32 = 98.6$
$\quad\quad$ So, 37°C $= 98.6°F$.

Find the equivalent Fahrenheit temperature. Use a calculator.

1. 20°C = _____ °F   **2.** 35°C = _____ °F

3. 85°C = _____ °F   **4.** 0°C = _____ °F

5. 56°C = _____ °F   **6.** 100°C = _____ °F

7. The temperature at the center of the sun is believed to be about 40 million degrees Celsius. What is the equivalent temperature in °F?

_____

8. Pedro, your Bolivian pen pal, reports that the temperature is about 25°C. Does Pedro need to wear a jacket outside?

_____

9. John wants to play baseball. It's sunny and 20°C. Is it a good day for baseball?

_____

Parts of Wholes!

Each picture shows a fraction of a whole number. Each shape represents a whole. Write the number sentence for each picture.

1.

_____

2.

_____

3.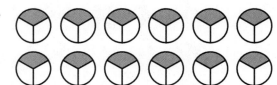

_____

4.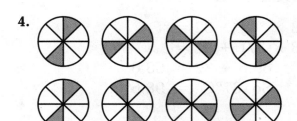

_____

5.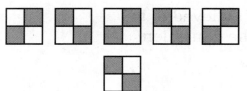

_____

6.

_____

3-D Fractions

You can use a cube to create a 3-dimensional model to multiply $\frac{1}{4} \times \frac{1}{3} \times \frac{1}{2}$.

Show $\frac{1}{4}$.	Show $\frac{1}{3}$.	Show $\frac{1}{2}$.
Cut the width into 4 equal parts.	Cut the height into 3 equal parts.	Cut the length into 2 equal parts.
Color 1 part blue.	Color 1 part red.	Color 1 part yellow.

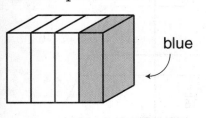

blue

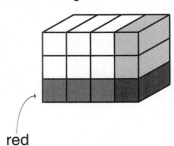

red

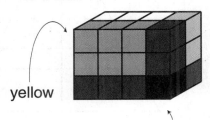

yellow

Lower-right front block has all 3 colors.

There are 24 blocks in the cube. There is only 1 block with all 3 colors.

So, $\frac{1}{4} \times \frac{1}{3} \times \frac{1}{2} = \frac{1}{24}$.

Color the 3-dimensional model to find the product.

1. $\frac{1}{3} \times \frac{1}{2} \times \frac{1}{2} = $ _____

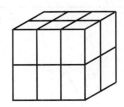

2. $\frac{1}{4} \times \frac{1}{2} \times \frac{1}{2} = $ _____

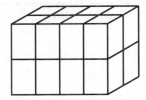

3. $\frac{1}{3} \times \frac{1}{3} \times \frac{1}{3} = $ _____

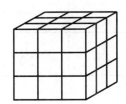

4. $\frac{1}{4} \times \frac{2}{3} \times \frac{1}{2} = $ _____

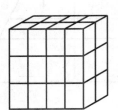

Fraction Triangles

Each triangle contains two multiplication sentences. Fill in the
missing numbers to complete each multiplication sentence.
Write your fractions in simplest terms.

1.

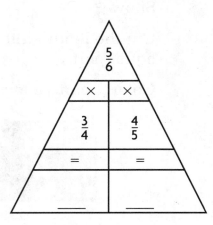

2.

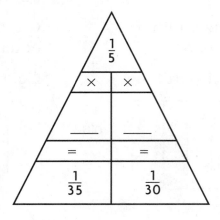

3.

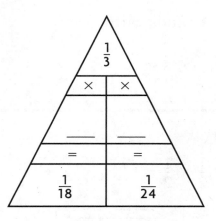

4.

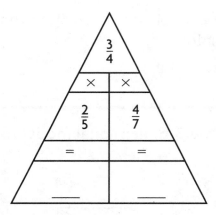

5.

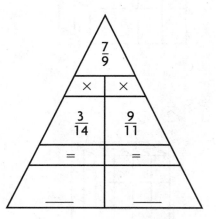

6.

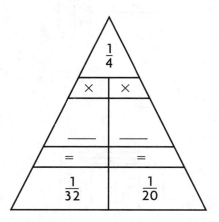

Four Square

Multiply in all four directions in each square. Record the products, in simplest form, in the circles.

1.

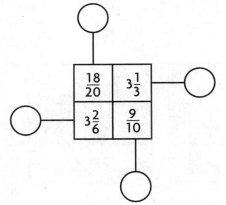

2.

3.

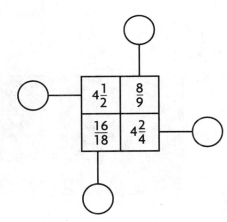

4.

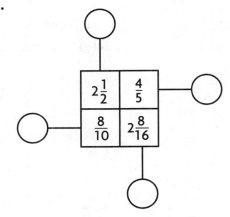

Which Model?

In each exercise, the fraction squares model a multiplication sentence. Choose a fraction and a mixed number from the list above the row of fraction squares. Write the multiplication sentence that is modeled.

1. $2\frac{1}{2}$ $\frac{3}{4}$ $2\frac{1}{5}$ $\frac{3}{5}$ $1\frac{7}{20}$ $2\frac{3}{5}$ $2\frac{1}{4}$

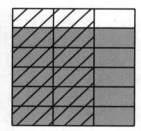

_____

2. $\frac{1}{3}$ $2\frac{5}{9}$ $\frac{3}{6}$ $3\frac{5}{6}$ $\frac{2}{3}$ $3\frac{5}{9}$ $\frac{5}{6}$

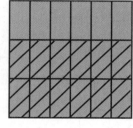
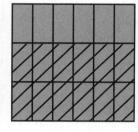

_____

3. Choose a mixed number from the list in Exercise 1 and a fraction from the list in Exercise 2. Make a model to multiply.

_____

What's the Point?

Draw line segments from 1 to 2, 2 to 3, 3 to 4, and 4 to 1.

Draw line segments from 5 to 6, 6 to 7, 7 to 8, and 8 to 5.

Draw line segments from 9 to 10, 10 to 11, 11 to 12, and 12 to 9.

Continue in this pattern until all the numbers have been used.

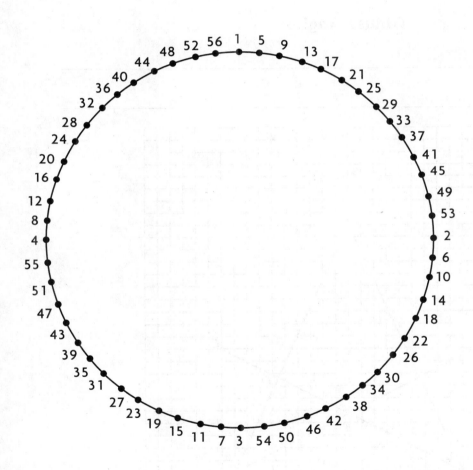

What shape appears in the center?

_____

A Star Is Born!

Name every acute angle and obtuse angle in the star.

Acute Angles

_____

_____

_____

Obtuse Angles

_____

_____

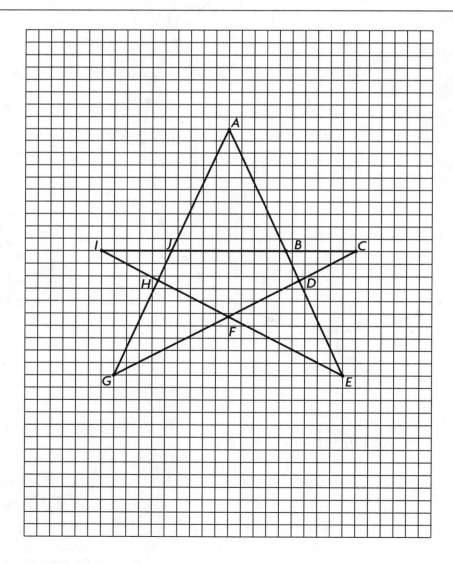

Divide and Conquer!

The enemy forces are upon you. They are rapidly approaching in the shape of a wedge. Your goal is to divide them by finding a line between their advancing forces. Study the step-by-step process for bisecting an angle to accomplish your goal.

Step 1

Draw an angle. Place the compass on the vertex, M. Draw an arc.

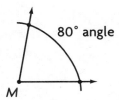

Step 2

Label the points where the arc intersects the angle as points L and N.

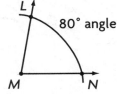

Step 3

Place the point of the compass on point L. Draw an arc with the pencil toward point N.

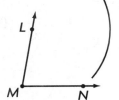

Step 4

Then place the point of the compass on point N. Draw another arc with the pencil to intersect the arc drawn from point L.

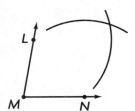

Step 5

Use a straightedge to draw a line from the vertex, M, through the two arcs at the point where they intersect.

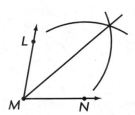

Bisect each angle.

1.

2.

The Path of Least Resistance

You must cross the Land of Hidden Angles. Make a path on
the map below to show the best direction from start to finish.
Include degrees of angles that you turn and number of miles
you walk for each section of your journey.

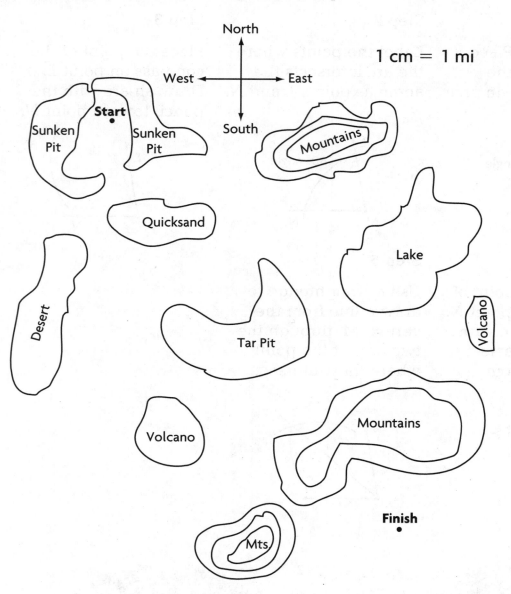

1 cm = 1 mi

How many miles in all did you walk?

_____

Venn Diagrams

Venn diagrams are used to show how objects are related to each other.

In a group of 10 pet owners, 5 own fish and birds, 2 own fish only, and 3 own birds only.

In the Venn diagram, the box represents all 10 pet owners. The area where the two circles overlap represents pet owners who own both fish and birds.

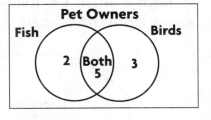

In another group of 10 pet owners, 6 own horses and 4 own pigs.

In the Venn diagram, the box represents all 10 pet owners. The circles do not overlap because no pet owner owns both horses and pigs.

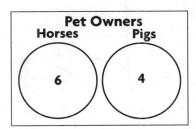

Use the definitions of the different quadrilaterals to construct a Venn diagram.

Trapezoid	Parallelogram	Rectangle	Rhombus	Square
1 pair of parallel sides	2 pairs of congruent sides 2 pairs of parallel sides	2 pairs of congruent sides 4 congruent angles 2 pairs of parallel sides	4 congruent sides 2 pairs of congruent angles 2 pairs of parallel sides	4 congruent sides 4 congruent angles 2 pairs of parallel sides

Quadrilaterals

Triangle Land

Triangle Land is divided into three counties. They are Isosceles County, Scalene County, and Equilateral County. Measure the sides of each triangle. Decide which county it belongs in. Then draw an arrow from the triangle to the correct county.

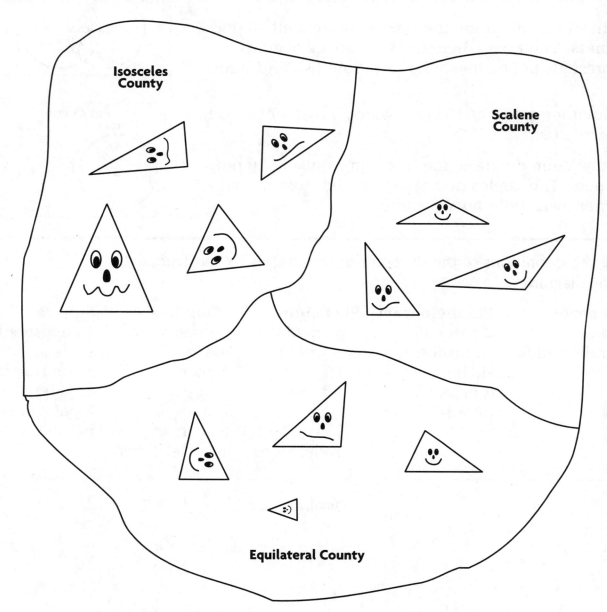

How many angles are there in each county?

_____

_____

Measure Me

On the planet Trianglium, there are three different species of triangles. They are Obtusiums, Acutiums, and Rightiums. Obtusiums have one obtuse angle and never smile. Acutiums always smile and have all acute angles. Rightiums feel they are always right because they have a right angle. They always smirk. The one thing they all have in common is that the sum of their angles always adds up to 180°.

Name each creature. Write *Obtusium, Acutium,* or *Rightium,* and draw the matching faces. Measure each angle of the triangle.

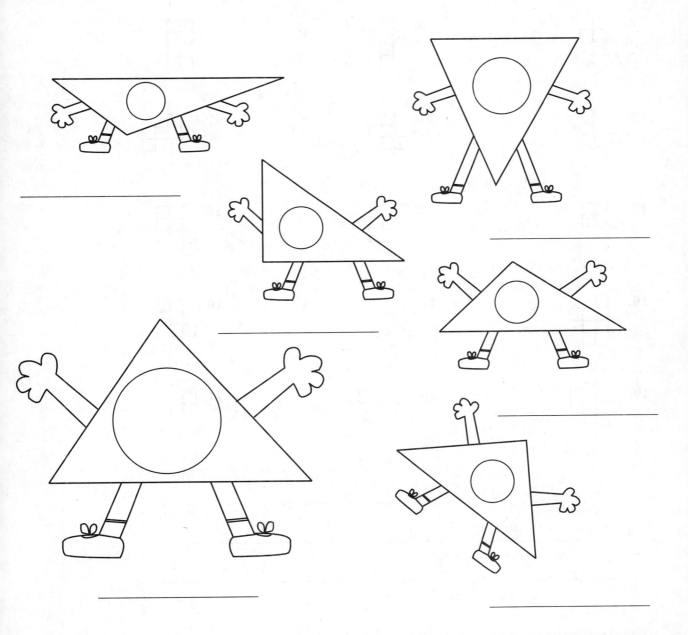

Name _____

Find the Congruent Shapes!

List below all the shapes that are congruent to each other.

1.

2.

3.

4.

5.

6.

7.

8.

9.

10.

11.

12.

13.

14.

15.

16.

17.

18.

19.

20.

21.

_____

_____

_____

Finish the Pictures

Each figure has one or two lines of symmetry. Only part of the
figure is drawn. Use the lines of symmetry to finish drawing
each picture.

1.

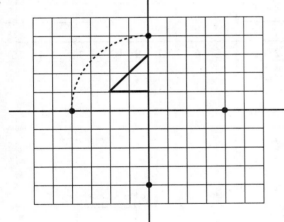

2.

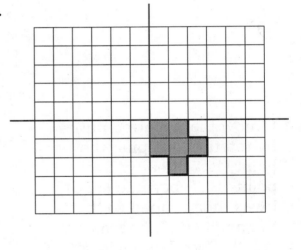

3.

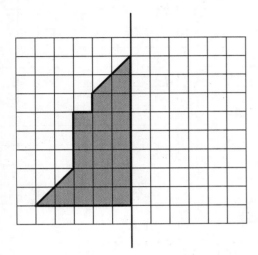

4.

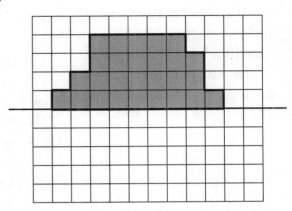

5.

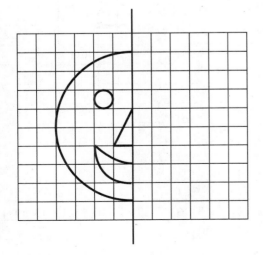

6.

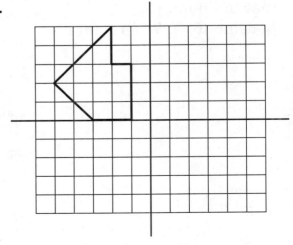

How Did I Get Here?

Use the clues to trace the movements of each figure. Draw the
figure as it is transformed from one position to the next.

1. I know that I was flipped
   so that my vertex at (4,1)
   went to (6,1). I was then
   slid 3 up and 3 to the right.
   Finally, I was rotated 90°
   to the right around my
   bottom right point. Where
   am I?

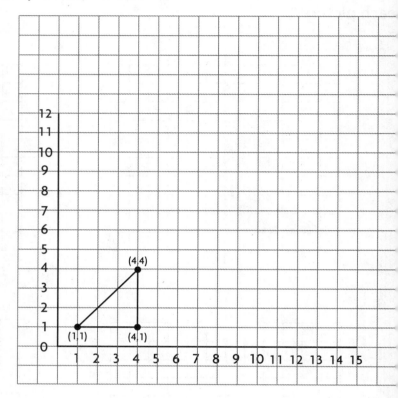

2. I know that I started out
   by rotating, keeping the
   point (12,7) steady. The
   point (10,7) became (12,5).
   I was then flipped so that
   the point (12,5) became
   (12,2). Then I was slid so
   that the point (12,2)
   became (6,4). Where am I?

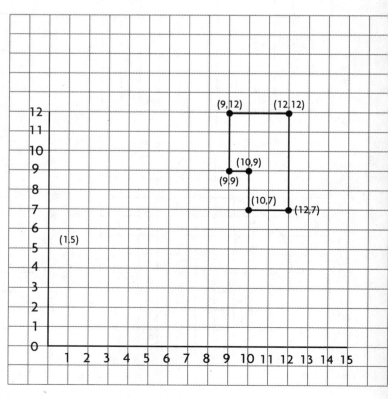

Other Tessellations

You can create figures that tessellate by changing the figures that you know tessellate.

The rectangle below tessellates.
Cut out two triangles from one side.

Attach the two triangles to the opposite side. Trace your new figure to show that it tessellates.

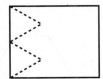

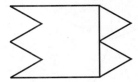

You can also create figures of different creatures.

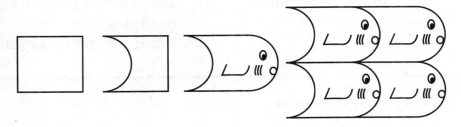

Trace and cut out each figure. Change it in the manner described above to create a new figure. Trace it to show that it tessellates.

1.

2.

3.

4. Pick one of the above figures and create an amusing creature.

What Tessellates!

You can classify tessellations using a code based on the
number of sides of each shape. Start with the smallest shape.
Then go clockwise around a point where all the shapes meet.

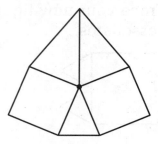

This is a 3.4.3.3.4.
The 3's stand for the
triangles, and the 4's
stand for the squares.

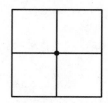

This is a 4.4.4.4.
The 4's stand for the
four squares.

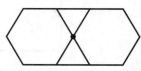

This is a 3.6.3.6.
The 3's stand for the
triangles, and the 6's
stand for the hexagons.

Draw what each classification would look like.

1. 3.3.3.4.4

2. 3.3.3.3.6

3. 3.4.6.4

4. 3.3.6.6.3.3

5. 3.4.4.6

6. 4.8.4.8

Blowing Up Balloons!

Sally and her friends are having a party. They need your help decorating balloons. Use the table below, a compass, and a centimeter ruler to construct and color the balloons for Sally and her friends. Then fill in the table with the diameter of each balloon.

Balloon Number	Radius	Diameter	Color
1	1 cm	_____	red
2	3 cm	_____	blue
3	2 cm	_____	orange
4	1.5 cm	_____	green
5	1 cm	_____	yellow
6	1 cm	_____	purple

1.

2.

3.

4.

5.

6.

Can You Get Around This?

Look at each object. Use a calculator to find its circumference.
Round to the nearest hundredth. Then complete the table below.

Remember, Circumference (C) = diameter × 3.14.

1.

Diameter of wheel = 24 in.

C = _____

2.

Diameter of mug = 8 in.

C = _____

3.

Diameter of coins = 2 in.

C = _____

4.

Diameter of thermometer = 0.75 in.

C = _____

5. Which object has the greatest circumference?

_____

6. What do you notice about the relationship of the diameter to the circumference in the objects above?

_____

_____

_____

Object	Circumference
_____	6.28 in.
_____	2.36 in.
_____	25.12 in.
_____	75.36 in.

Superstar Power!

Pick the answer that is more reasonable as the circumference of each item. Match the letter or symbol of each item with the measurements in the box at the bottom of the page to find the secret message.

C = circumference

A 1 cm or 1 m   **Y** 3 cm or 3 dm   **S** 0.5 cm or 5 cm   **T** 2.5 cm or 2.5 m

candle

banana

pencil

power
pole

C = _____   C = _____   C = _____   C = _____

E 2 cm or 2 m   **N** 25 cm or 2.5 cm   **O** 23 cm or 23 m   **P** 32 cm or 3.2 cm

tube

glass

jar

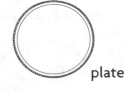
plate

C = _____   C = _____   C = _____   C = _____

R 2.5 cm or .25 cm   **U** 2.8 cm or 28 cm   **D** 4.7 cm or 47 cm   **!** 1.2 cm or 1.2 m

hot dog

jar lid

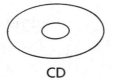
CD

globe

C = _____   C = _____   C = _____   C = _____

___	___	___			___	___	___
3 cm	23 cm	28 cm			1 cm	2.5 cm	2 cm
___			___	___	___	___	___
1 cm			0.5 cm	28 cm	32 cm	2 cm	2.5 cm
___	___	___	___	___	___	___	___
0.5 cm	2.5 m	28 cm	47 cm	2 cm	25 cm	2.5 m	1.2 m

Name _____

Wanted—Missing Angles!

Ryan is on a mission. Help him find each angle. Look at the circles below and write in the degrees of the missing angles. Then use the table below to color the angles you find.

1.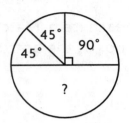

angle = _____

2.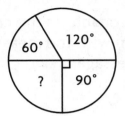

angle = _____

3.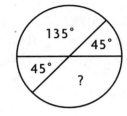

angle = _____

4.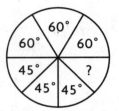

angle = _____

5.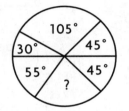

angle = _____

6.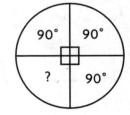

angle = _____

7.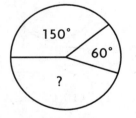

angle = _____

8.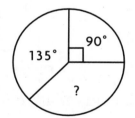

angle = _____

9.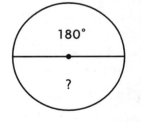

angle = _____

10.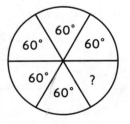

angle = _____

11.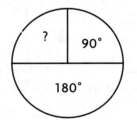

angle = _____

12.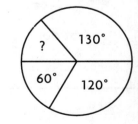

angle = _____

Angle	Color	Angle	Color	Angle	Color	Angle	Color
60°	green	90°	red	50°	orange	135°	brown
45°	yellow	180°	blue	150°	pink	80°	purple

Paula's Pies

Paula sells her home-baked pies by the slice. She cuts her pies in 10 slices of equal size. This means each slice has a measure of 36°.

Use a protractor to find out how many slices Paula has sold of each pie.

1.

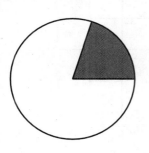

_____

2.

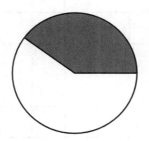

_____

3.

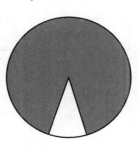

_____

4.

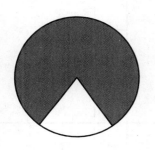

_____

5.

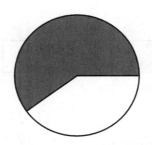

_____

6.

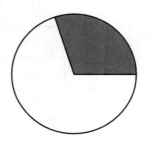

_____

7.

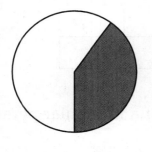

_____

8.

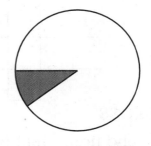

_____

9.

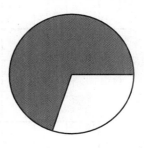

_____

10. If Paula sells 180° of a pie, how many slices are

left? _____

11. Each pie above has 2 angles. Name the pairs of pies that have the same 2 angles.

_____

_____

Crossed Words

ACROSS

2. solid figure that has two congruent polygons in parallel planes

5. All 6 faces of this solid figure are
   rectangles.

6. where two faces of a solid figure
   meet

7. The base of this solid figure
   has 4 equal sides. The faces are
   triangles.

9. five-sided polygon

10. formed where three or more
    edges of a solid figure meet

11. The faces of a triangular pyramid
    are these.

12. solid figure with triangular bases
    and rectangular faces

DOWN

1. flat surface of a solid figure

2. solid figure with a pentagon for a
   base and triangular faces

3. solid figure with one polygon
   base and triangular faces

4. congruent polygons in parallel
   planes

8. These form the bases of a prism.

Create a Package

Describe and then design a net that could be used to wrap each of the following gifts. Use crayons or markers to create a colorful pattern on your net.

1.

_____

2.

_____

3.

_____

4.

_____

What a View!

Some everyday objects are sketched below. Draw a top view,
a side view, and a front view of each.

Top view	Side view	Front view

1.

2.

3.

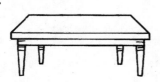

4.

Number Patterns

Find the next three terms. Write the rule used to form the pattern.

1. 27, 39, 63, 99, _____

Rule: _____

_____

2. 663, 638, 588, 513, _____

Rule: _____

_____

3. 117, 104, 135, 122, 153, _____

Rule: _____

_____

4. 84; 336; 168; 672; 336; _____

Rule: _____

_____

5. 189, 235, 282, 330, _____

Rule: _____

_____

6. 79; 474; 158; 948; 316; _____

Rule: _____

_____

Operation 45

Each circle holds four numbers. If you perform three different operations on these numbers, using each number once, the result is 45.

Write the three number sentences on the lines. The first one has been done for you!

1.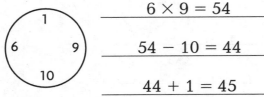

$$6 \times 9 = 54$$

$$54 - 10 = 44$$

$$44 + 1 = 45$$

2.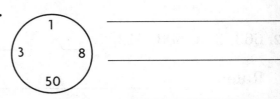

_____

_____

_____

3.

_____

_____

_____

4.

_____

_____

_____

5.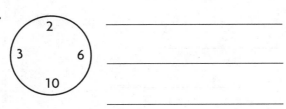

_____

_____

_____

6.

_____

_____

_____

Cube Counting

Each cube arrangement contains only one shaded cube. Carefully observe each cube arrangement. Then answer the questions below.

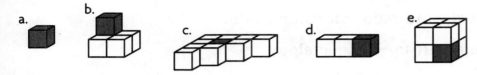

a.   b.   c.   d.   e.

1. What fractional part of Figure b is represented by the shaded cube?

_____

2. What fractional part of Figure d is represented by the shaded cube?

_____

3. What fractional part of Figure e is represented by the white cubes?

_____

4. What decimal represents the part of Figure c made up by the shaded cube?

_____

5. What decimal represents the part of Figure b made up by the white cubes?

_____

6. What decimal represents the part of Figure a made up by the shaded cube?

_____

7. Identify the part of a figure that could be represented by the fraction $\frac{2}{3}$.

_____

8. Identify the part of a figure that could be represented by the decimal 0.9.

_____

Word Ratios

Just as ratios can be used to compare numbers, analogies can be used to compare words.

An analogy sets up two word ratios for comparison.

Here's an example of a part-to-whole analogy.

duck : bird = poodle : dog

This analogy is read "duck is to bird as poodle is to dog."

Remember that ratios must be in the same order.

duck : bird = dog : poodle ⟵——— Incorrect

Circle the letter that best completes each analogy.

1. general : army

    a. gardener : garden

    b. teacher : class

    c. school : principal

    d. chef : kitchen

2. water : ocean

    a. dam : river

    b. Jupiter : solar system

    c. ocean : sea

    d. sand : beach

3. taco : dinner

    a. cream cheese : bagel

    b. donut : breakfast

    c. dessert : ice cream

    d. popcorn : movie

4. Houston : Texas

    a. Berlin : Germany

    b. Mississippi : state

    c. Nevada : Las Vegas

    d. mountain : Everest

5. pianist : musician

    a. bird : feathered

    b. boat : canoe

    c. ant : insect

    d. airplane : flight

6. Frankenstein : monster

    a. Wolfman : full moon

    b. Dracula : vampire

    c. Buffy : vampire slayer

    d. mummy : Eygpt

Winning Ratios

Contest promoters are often required to tell you your chance of winning.

Here, the chance of winning is expressed as the ratio 1:100,000. This means that, for every 100,000 people who enter the contest, there is one winner.

> # Enter Contest!
> # Win a Fabulous Prize!
> Chance of winning 1:100,000

The following data concern a cereal-box contest.

Prize	Value of Prize	Chance of Winning
Grand Prize	$20,000	1:2,500,000
Second Prize	$300	1:42,000
Third Prize	$20	1:3,000
Fourth Prize	$1	1:50

Use the data about the cereal-box contest to solve each problem.

1. Which prize is awarded most often? least often?

_____

2. What is your chance of winning second prize? _____

3. What is your chance of winning a $20 prize? _____

4. If everyone in your town or city entered the contest, what

results would you expect? _____

_____

5. The promoters of the contest advertise that the chance of winning a prize is "better than 1 in 50." Is this accurate? Is it misleading?

Explain. _____

_____

Ratio Art

You can use ratios to draw the same piece of art in different sizes. Redraw the following piece of art on the grid below using the ratios 1:2, 1:3, and 1:4.

Color in your drawings.

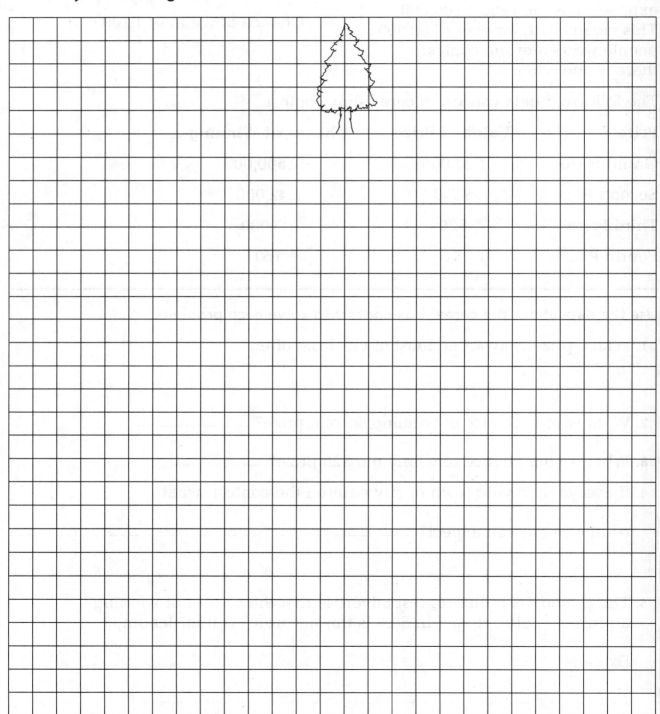

Scavenger Hunt

Ratios are everywhere in real life. Here are some examples.

- The scale on a road map reads: $\frac{1}{2}$ in. = 20 mi.

- A baseball player has 2 hits in 5 at-bats.

- A video 10-pack is priced at $19.95.

How many ratios can you find? Look for ratios at the store, at school, at home, in magazines, and in newspapers. Record the ratios you find below, and tell where you found them.

Ratio Found	Where Ratio Found
1. "Four out of five dentists recommend . . ."	TV ad
2. "4 servings per package"	spaghetti package
3.	
4.	
5.	
6.	
7.	
8.	
9.	
10.	

Rate the Ratios

Regulations state that the United States flag have a
width-to-length ratio of 1:1.9.

Find the length of each flag.

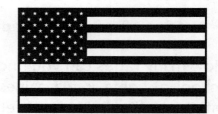

1. width = 3 yards, length = _____

2. width = 5 meters, length = _____

3. width = 12 feet, length = _____

4. width = 26 inches, length = _____

Ratios state the relationship between two quantities. For
a ratio to be in simplest form, both terms must have the
same units.

Find the ratios in simplest form.

5. 12 to 6 _____

6. 6 to 12 _____

7. 7 miles to 18 miles _____

8. 4 feet to 5 inches _____

9. $3\frac{1}{2}$ days to 10 hours _____

10. 144 ships to 18 ships _____

11. 3 pints to 9 quarts _____

12. 6 minutes to 20 seconds _____

13. 18 feet to 9 yards _____

14. a leap year to 61 days _____

15. 4 cups to 2 gallons _____

16. 1 mile to 440 yards _____

17. 4 feet to 76 feet _____

18. 2 weeks to 1 year _____

19. $30.00 to $0.75 _____

20. 2 years to 13 weeks _____

Solar Ratios

You've been assigned to help build a model of the solar system for a local park. Your job is to figure out how far from the sun to place the planets. Here's some useful information.

Distance from Sun to Each Planet	
Mercury—36,000,000 mi	Saturn—886,400,000 mi
Venus—67,000,000 mi	Uranus—1,786,000,000 mi
Earth—93,000,000 mi	Neptune—2,794,000,000 mi
Mars—141,000,000 mi	Pluto—3,660,000,000 mi
Jupiter—483,300,000 mi	

Since the sculptor wants the Earth to be 10 feet from the sun, you need to convert the planet distances into Earth units, then multiply by 10.

$$\text{Earth units} = \frac{\text{planet's distance from the sun}}{\text{Earth's distance from the sun}}$$

$$\text{Earth units for Mercury} = \frac{36,000,000 \text{ miles}}{93,000,000 \text{ miles}} = 0.387$$

Mercury's distance in model = $0.387 \times 10 = 3.87$ feet

Complete the chart.

Planet	Distance in Earth Units	Distance in Model
Mercury	0.387 ft	3.87 ft
Venus	0.720 ft	7.2 ft
Earth	1.000 ft	10 ft
Mars		
Jupiter		
Saturn		
Uranus		
Neptune		
Pluto		

Sale!!

Storewide Sale!!

Pilar is a clerk at a women's clothing store. Her boss wants her to mark each price tag with the sale price. Can you help her?

On each tag, cross off the regular price and write in the sale price.

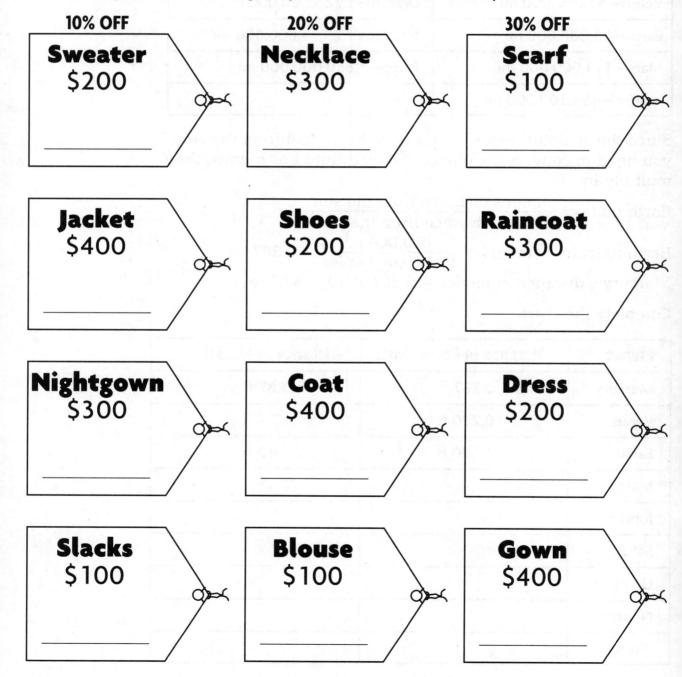

10% OFF

Sweater
$200

_____

Jacket
$400

_____

Nightgown
$300

_____

Slacks
$100

_____

20% OFF

Necklace
$300

_____

Shoes
$200

_____

Coat
$400

_____

Blouse
$100

_____

30% OFF

Scarf
$100

_____

Raincoat
$300

_____

Dress
$200

_____

Gown
$400

_____

Understanding Standings

Check the sports section of your newspaper. Teams are ranked according to "winning percentage." This may be abbreviated as "pct." The team with the highest winning percentage is in first place.

To find a team's winning percentage, use your calculator. Here is an example that shows you how.

Suppose the Rams won 17 and lost 11. That's 28 games played in all.

pct = wins ÷ games = 17 $\boxed{÷}$ 28 $\boxed{=}$ $\boxed{0.6071428}$

Winning percentage is usually expressed as a decimal rounded to three decimal places.

So, the Rams' pct is 0.607, or .607.

Rank the following teams from highest to lowest winning percentages. Their records are given. Use your calculator to find each winning percentage to three decimal places.

	Wins	Losses	Pct	Rank
Cubs	34	20	_____	_____
Rams	22	29	_____	_____
Timberwolves	25	26	_____	_____
Dodgers	27	24	_____	_____
White Sox	22	31	_____	_____
Giants	28	25	_____	_____
Cardinals	20	23	_____	_____
Pacers	27	27	_____	_____

Find the Match!

Write the letter of the percent that matches each fraction.
Then place the letter above each exercise number below to
solve the riddle.

1. _____ $\dfrac{17}{100}$ **B.** 76%   **Q.** 30%

2. _____ $\dfrac{3}{10}$

3. _____ $\dfrac{19}{25}$ **D.** 22%   **R.** 20%

4. _____ $\dfrac{1}{2}$

5. _____ $\dfrac{1}{5}$ **F.** 17%   **S.** 4%

6. _____ $\dfrac{1}{25}$

7. _____ $\dfrac{11}{50}$ **H.** 50%   **T.** 35%

8. _____ $\dfrac{7}{20}$

9. _____ $\dfrac{12}{25}$ **I.** 3%   **V.** 48%

10. _____ $\dfrac{7}{10}$

11. _____ $\dfrac{81}{100}$ **K.** 7%   **W.** 81%

12. _____ $\dfrac{3}{4}$

13. _____ $\dfrac{7}{100}$ **M.** 40%   **X.** 70%

14. _____ $\dfrac{3}{100}$

15. _____ $\dfrac{2}{5}$ **O.** 75%

What must have four wings to fly?

___ ___ ___   ___ ___ ___ ___ ___
 8   11   12   3   14   5   7   6

Name _____

Percents with Style

There are many different ways you can shade a 10-by-10 grid
to show a percent. Here are three different models for 24%.
The first model is familiar. The second and third models
are fun.

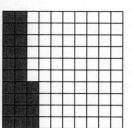 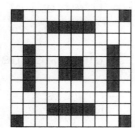 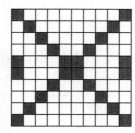

Show each percent by shading an interesting pattern.

1. 36%

2. 80%

3. 40%

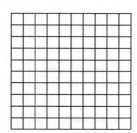 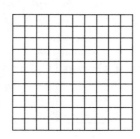 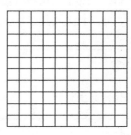

4. 28%

5. 60%

6. 72%

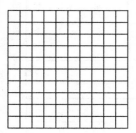 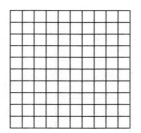 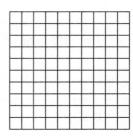

7. 18%

8. 44%

9. 66%

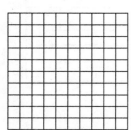 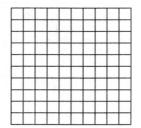 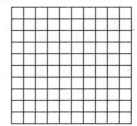

Fast-Food Facts

What percent of the calories you consume are from fat? If you want to keep fat below 30%, go light on the fast food!

To calculate the percent of calories from fat, use a calculator.

Calories from fat $\boxed{\div}$ total calories $\boxed{\times}$ 100 $\boxed{=}$ percent of calories from fat.

Suppose a slice of bread has 90 calories, and 12 calories come from fat.

$$12 \boxed{\div} 90 \boxed{\times} 100 \boxed{=} \boxed{13.33333}$$

So, about 13% of the calories in this bread come from fat.

For each of the following fast foods, calculate the percent of calories from fat. Use your calculator.

Fast Foods	Total Calories	Calories from Fat	Percent Fat
Harry's Hamburger	420	180	_____
Harry's Hamburger with Cheese	520	260	_____
Big Bun Burger	640	350	_____
Big Bun Burger with Cheese	730	410	_____
Teresa's Taco	180	100	_____
Teresa's Taco Light	140	50	_____
Charlie's Cheese Pizza (slice)	344	90	_____
Bob's Big Bacon Burger	610	290	_____
Crispy Chicky	360	180	_____

Pascal's Triangle

This triangular arrangement of numbers is named for Blaise Pascal, a seventeenth-century mathematician.

1. Find the pattern in each row to complete the triangle.

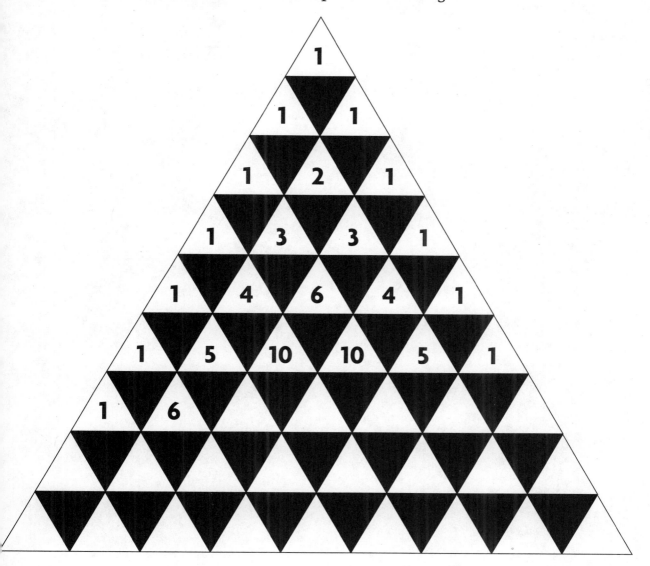

Complete.

2. The number that begins and ends each horizontal row is _____.

3. Each number is the _____ of the two numbers above it.

4. Nine rows are shown. Write the numbers that would be in the tenth row.

_____